Cybersecurity and Privacy –
Bridging the Gap

WIRELESS WORLD
RESEARCH FORUM

RIVER PUBLISHERS SERIES IN COMMUNICATIONS

Series Editors

ABBAS JAMALIPOUR
The University of Sydney
Australia

MARINA RUGGIERI
University of Rome Tor Vergata
Italy

Indexing: All books published in this series are submitted to Thomson Reuters Book Citation Index (BkCI), CrossRef and to Google Scholar.

The "River Publishers Series in Communications" is a series of comprehensive academic and professional books which focus on communication and network systems. The series focuses on topics ranging from the theory and use of systems involving all terminals, computers, and information processors; wired and wireless networks; and network layouts, protocols, architectures, and implementations. Furthermore, developments toward new market demands in systems, products, and technologies such as personal communications services, multimedia systems, enterprise networks, and optical communications systems are also covered.

Books published in the series include research monographs, edited volumes, handbooks and textbooks. The books provide professionals, researchers, educators, and advanced students in the field with an invaluable insight into the latest research and developments.

Topics covered in the series include, but are by no means restricted to the following:

- Wireless Communications
- Networks
- Security
- Antennas & Propagation
- Microwaves
- Software Defined Radio

For a list of other books in this series, visit www.riverpublishers.com

Cybersecurity and Privacy – Bridging the Gap

Editors

Samant Khajuria
Lene Tolstrup Sørensen
Knud Erik Skouby

CMI/Aalborg University
Denmark

River Publishers

Published, sold and distributed by:
River Publishers
Alsbjergvej 10
9260 Gistrup
Denmark

River Publishers
Lange Geer 44
2611 PW Delft
The Netherlands

Tel.: +45369953197
www.riverpublishers.com

ISBN: 978-87-93519-66-4 (Hardback)
       978-87-93519-65-7 (Ebook)

©2017 River Publishers

Contents

Foreword

The WWRF Series in Mobile Telecommunications

The Wireless World Research Forum (WWRF) is a global organization bringing together researchers into a wide range of aspects of mobile and wireless communications, from industry and academia, to identify the key research challenges and opportunities. Members and meeting participants work together to present their research and develop white papers and other publications on the way to the Wireless World. Much more information on the Forum, and details of its publication programme, are available on the WWRF website www.wwrf.ch. The scope of WWRF includes not just the study of novel radio technologies and the development of the core network, but also the way in which applications and services are developed, and the investigation of how to meet user needs and requirements.

WWRF's publication programme includes use of social media, online publication via our website and special issues of well-respected journals. In addition, where we have identified significant deserving subjects, WWRF is keen to support the publication of extended expositions of our material in book form, either singly-authored or bringing together contributions from a number of authors. This series, published by River Publications, is focused on treating important concepts in some depth and bringing them to a wide readership in a timely way. Some will be based on extending existing white papers, while others are based on the output from WWRF-sponsored events or from proposals from individual members.

We believe that each volume of this series will be useful and informative to its readership, and will also contribute to further debate and contributions to WWRF and more widely.

Dr. Nigel Jefferies
WWRF Chairman

Professor Klaus David
WWRF Publications Chair

Preface

This book is motivated by activities and collaborations within World Research Wireless Forum (WWRF) especially Working Group A/B. Cybersecurity and privacy are issues that in a connected world are raising concerns of wide groups including the scientific community; industry; governments and politicians; and civic groups. There are daily reports on major security breaches with huge economic and social impact and on inappropriate use of private data. This result in loss of trust in the digital platforms companies loose revenue and public institutions are discredited.

Global trends in cybersecurity and privacy are discussed and possible remedies to counter unwarranted developments are analyzed from both academic and industry point of views.

The book is addressing security professionals and university audiences.

Sincerely thanks to authors Anurag Bihani, STES's Sinhgad Institute of Technology and Science, Pune, India; Aske Hornbæk Knudsen, Electronic Systems, Aalborg University, Denmark; Bipjeet Kaur, Technical University of Denmark, Ballerup Campus (DTU); BirgerAndersen, Technical University of Denmark, Ballerup Campus (DTU); Charles A. Clarke, School of Computer Science and Mathematics, Kingston University, Kingston upon Thames, London, United Kingdom; DimitrisTsaptsinos, School of Computer Science and Mathematics, Kingston University, Kingston upon Thames, London, United Kingdom; Eckhard Pfluegel, School of Computer Science and Mathematics, Kingston University, Kingston upon Thames, London, United Kingdom; Egon Kidmose, Electronic Systems, Aalborg University, Denmark; Fei Liu, 2012 Shield Laboratories, Huawei Technologies, Singapore; Greig Paul, Department of Electronic & Electrical Engineering, University of Strathclyde, Glasgow, United Kingdom; Henning Olesen, Center for Communication, Media and Information Technologies (CMI), Aalborg University Copenhagen, Copenhagen, Denmark; Henrik Tange, Technical University of Denmark, Ballerup Campus (DTU); James Irvine, Department of Electronic & Electrical Engineering, University of Strathclyde, Glasgow,

United Kingdom; James Orwell, School of Computer Science and Mathematics, Kingston; University, Kingston upon Thames, London, United Kingdom; Jens Myrup Pedersen, Electronic Systems, Aalborg University, Denmark; Joakim G. Randulff, School of Computer Science and Mathematics, Kingston University, Kingston upon Thames, London, United Kingdom; Kristoffer Rohde, Principal Sales Engineer, Opentext/EMC/Dell; Lars Kierkegaard, Founder of InnoPaze; Marcus Wong, Wireless Security Research and Standardization, Huawei Technologies, USA; Mikki Alexander, Electronic Systems, Aalborg University, Denmark; Mousing Sørensen, Electronic Systems, Aalborg University, Denmark; Ole Kjeldsen, Microsoft Corporation, Kongens Lyngby, Denmark; Prashant S. Dhotre, Center for Communication, Media and Information Technologies (CMI), Aalborg University Copenhagen,Denmark; Samant Khajuria, Center for Communication, Media and Information Technologies (CMI), Aalborg University Copenhagen, Denmark; Theis Dahl Villumsen, Electronic Systems, Aalborg University, Denmark

We are grateful for inspiration and support from the publishers and WWRF.

Finally, but not least thanks to Samant Khajuria Lene Tolstrup Sørensen for getting the idea of the book and for tireless effort in collecting the contributions and editing.

Knud Erik Skouby
Copenhagen, March 2017

List of Figures

List of Tables

xix

List of Abbreviations

AM	Amplitude Modulation
ANSI	American National Standards Institute
DAB	Digital Audio Broadcast
DECT	Digital Enhanced Cordless Telecommunications
DTT	Digital Terrestrial Television
DVB-T	Digital Video Broadcast-Terrestrial
FM	Frequency Modulation
GDPR	General Data Protection Regulation
GSM	Global System for Mobile communication
ICT	Information and Communication Technology
ISO	International Standards Organisation
LTE	Long Term Evolution
NOC	Network Operation Centre
UMTS	Universal Mobile Telecommunication System
UPS	Uninterruptable Power Supply
WAN	Wide Area Network

Introduction

Cybersecurity and privacy are global issues with a local impact. When any security or privacy breach occurs, it impacts our society at all levels. Trust in the digital platforms is lost; companies loose revenue and even public institutions are discredited. These two issues have become major concerns of wide groups of today's globally connected society; the scientific community; industry; governments and politicians; and civic groups.

Our every-day life is very much dependent on the connected devices we are surrounded by – mobile phones, computer in homes and offices, sensors and machines in factories, hospitals etc. Cybersecurity and privacy concerns are fueled by the collection of huge amount of data exchanged by the connected devices and increasingly collected by Internet of Things (IoT). The IoT technology integrates the Internet as we know it today into a multitude of things, and hence commonly known objects such as clothes, food packing toothbrushes, etc. In this integration a huge amount of data is created. It will be stored in data centers and accessible as via cloud services in cross referenced selections. This vision of IoT incorporates new forms of communication between people and things and between things themselves. A multiplicity of devices/sensors acts outside the reach of the humans: sensors and devices communicate via Internet, analyze and act upon the data they select to provide services for the user. This automated communication and action is a key-feature in the potential of Information and Communication Technologies (ICTs). The interconnected physical devices, vehicles; buildings, and other items embedded with electronics collect and exchange data as basis for new services effectuated on predefined conditions that, however, may be modified in the provision process. The current view is that there will be more than 30 billion interacting devices by the year 2020 (e.g., Forbes, 2016) and the IoT vision foresees use cases in ever more sensitive areas such as e-health, Intelligent Transport System, Smart Grids, smart homes (Patel et al., 2016). It is estimated that around 70% of these will have vulnerability against cybersecurity (Security Intelligence, 2016) as the sharing of data

1

through devices and smart-networks is everywhere and the data is stored in data centers and accessible as via cloud services in cross referenced selections.

The concern is that in the hyper-connected society cybersecurity or rather lack of it and potential privacy infringement challenge the realization of the potentials of new ICT based services. Related to this, the two concepts of cybersecurity and privacy are increasingly seen as complicated and interrelated.

In the past, cybersecurity used to be simple. Most of the cyber threats used to be random acts of vandalism by the means of viruses, worms and Trojans. With up-to-date firewalls and anti-viruses, users appeared to be protected and relatively safe. But today the richness of resources available in this ecosystem has made the environment very attractive to new forms of attacks leading to a new wave of risks where attackers have moved from random attacks to targeted attacks and behind every attack, attackers have different motivations and objectives. These attackers are broadly classified into five categories: Commodity threats, Hacktivists, Organized crime or cybercrime, espionage and cyberwar.

The complex and difficult field of cybersecurity is intrinsically inter-disciplinary. Policies and laws, objectives and technologies, strategies and procedures all of these topics play a role in cybersecurity and all of these are tightly interconnected with activities in one field immediately affecting the others.

The concept of privacy dates back to Westin (1968) who defined the term Information privacy as: "the right to select what personal information is known to what people" (Westin, 1968). However, this definition was developed during a time when digital services not existed and definitions that take this into account must be used today. Ziegeldorf et al. (2013) defines privacy in an IoT intra-structure as: "Privacy in the Internet of Things is the threefold guarantee to the subject for awareness of privacy risks imposed by smart things and services surrounding the data subject; individual control over the collection and processing of personal information by the surrounding smart things; and awareness and control of subsequent use and dissemination of personal information by those entities to any entity outside the subject's personal control space".

These definitions imply that the user must be in control of the collection and processing of personal information and at the same time must be aware and take control. With the IoT infra-structure, the individual control of private data is basically impossible. The lack of transparency, the selling of data to third parties and the fundamental lack of understanding of individual users complicates the management of privacy; certainly for individuals but increasingly also for states/governments. One response of governments is to have data of their citizens kept onshore where some regulation can be enforced. A number of governments have passed or passing cybersecurity laws including also Russia, Indonesia and China (Financial times 1 March, 2017). One of the most elaborate systems is the EU's General Data Protection Regulation (GDPR) with the aim to build or increase trust in EU citizens in using digital services on the Digital Single Market. With a fully functioning Digital Single Market, the union claims to create up to 415 billion euro in revenue and hundreds of thousands of new job. But today citizens do not trust online services, which mean they will not use all the opportunities presented by technologies. GDPR is seen as a modernization of the protection of the processing of personal data to cover the legislative gaps created by the rise of social media, big data and increasingly digital world.

This book analyses and discusses ongoing technical, legal and socioeconomic trends seen from academia and industry providing a background for understanding the dynamic and interrelated concepts cybersecurity and privacy that are shaping all aspects of our future society.

Contributions – from Academia

The chapter titled *"An Introduction to Security Challenges in User-Facing Cryptographic Software"* by G. Paul & J. Irvine discusses the trade-off between usability and security when developing secure software. For a good security application, the authors identified two main challenges.

- Ensure that the underlying implementation is technically secure, with correct and proper implementation of cryptographic techniques and resistant to attack
- Ensure that the software is suitably easy and intuitive for users to operate, while doing so in a secure manner.

The authors identified this trade-off especially in smartphone application to ensure good performance in mobile devices. Nine android encryption applications were selected for investigation, where 5 of them are published on the google play store by "top developers". Most of these applications are used for secure storage of multimedia content like photos and videos and passwords. The application analyzed could potentially be in use by as many as 117 million users.

Analysis showed that most of applications claiming to encrypt photographs or videos were simply leaving the vast majority of the file intact, with only a small header of the file differing between the plaintext and ciphertext. In one of the case, the first 100 bytes of the file header was flipped and in other case, the first 10 bytes of JPEG file header was removed from the file and replaced with zero byte value. Other parts of the file remain in plaintext and the bytes taken from the header were stored in SQLite database on the shared storage areas, which allows any application to easily retrieve the original first ten bytes.

In this chapter, the authors have also made a number of recommendations to improve the security of applications. These include a good quality random number sources for all the key generation processes and the use of master key to encrypt the file in addition to a user credential-derived key for master key protection. Second recommendation is the choice of cipher along with the suitable mode of operation for the cipher. Section 4.6.3, discusses the choice of encryption algorithm and recommends Advanced Encryption Standard (AES) as a suitable cipher. The section also compares different modes of operation eg., Electronic code book (ECB), which introduces a number of vulnerabilities and Cipher block chaining (CBC) which prevents identification of patterns within the plaintext from analysis of the ciphertext. Third recommendation is the need for authenticated encryption to prevent tampering with plaintext by sophisticated attackers. Section 4.6.5 discusses the three constructs; Encrypt-then-MAC (ETM), MAC-then-Encrypt (MTE) and Encrypt-and-MAC (EAM).

The chapter titled *"Take it or Leave it": Effective Visualization of Privacy Policies* by Prashant S. Dhorte et al. discusses the two most important principles ie., "notice & choice". These privacy protection principles focus on informing the users that their personal information is gathered, used and probably shared. The focus of notice and choice is most often on long statements of privacy policies, which are complex, time-consuming to read and questionable to what extent they actually achieve their goal of information the users.

In this chapter the authors have taken the case of India and analyzed over 600 privacy policies of the popular websites in the country. Due to different structures of privacy policies presented in different formats the author designed a corpus consisting of 43,544 sentences considering all categories. Different classifiers were designed for identifying each category. This was done to enable

Multi-Category Text Classification, i.e. each sentence can be part of more than one category.

A privacy policy elucidator tool (PPET) is also proposed, which can analyze and visualize privacy policies and help the users understand the information collection, storage, processing and sharing policies regarding the user's personally identifiable and non-personally identifiable data. The semi-automatic tool is designed as a privacy awareness tool than a privacy protection tool. The goal of the tool is to ease the task of reading a privacy policy and make it more engaging to the user. The tool fetches the analyzed contents of the privacy policy and effectively visualizes them for easy reading. The visualization of privacy policies is divided into eight categories ie., Information collection, way of information collection, purpose of the information collection, information sharing, cookies, policy for children, information security and others. Under these eight categories, the policies are visualized in a donut chart which represents the distribution of sentences of the policies across different sections. The tool has also incorporated the website reputation obtained from "Web of Trust (WOT)", which works on a unique crowdsourcing approach, where reviews and ratings are collected from millions of users across the globe.

As recommendations, the authors also propose a standard template of privacy policies in India which can further enhance the user's privacy awareness and understanding of privacy policies. The template addresses elements like – accessibility and format, readability, objective and scope, data collection, use of personally identifiable information(PII), information sharing and disclosures, customer choice and consent, security measures, notification and user feedback.

The Chapter titled *"A Secure Channel Using Social Messaging for Distributed Low-Entropy Steganography"* by E. Pfluegel et al. discusses how a secure channel can be established between the two users of online social networks (OSNs). The main goal here is to provide end-to-end security, meaning that contents of the message are not able to be decrypted by any third party including the OSN providers eg., Facebook, Twitter, Google+, which could hypothetically take the role of the "man-in-the-middle" of the same name.

In the chapter authors contributions are twofold:

- First contribution is the proposal for a secure channel, achieving confidentiality through undetectable communication based on distributed low-entropy steganography using social messaging.

- Second contribution is the development and release of a prototype implementation, in the form of an Android mobile app, using on communication using multi-channel mobile instant messaging through Twitter and Gmail.

The authors propose an architecture which consists of a cryptographic protocol and an OSN system architecture. The cryptographic protocol uses suitably adapted secret sharing scheme with random shares. This approach is low-entropy steganography with improved payload capacity. For OSN system architecture, the system uses a distributed architecture, improving on the two-channel approach by achieving additional security and redundancy. To achieve undetectable communication, the solution uses n socially constrained and one additional, socially unconstrained channel. Some assumptions are made for the scheme to be operational and secure which are as follows – users have accounts with n online social networks, on each of which they are mutual friends. Both the uses have joint access to external file server, Online social network authentication mechanism is trusted. The individual Online social network do not co-operate ie., confidentiality of the user data is respected with regards to external entities.

Section 4, demonstrates the proof-of-concept. An android application is developed, the implementation uses Twitter and Gmail as online social network. An open source library is used to implement the secret sharing functionality, based on Shamir's secret sharing scheme. The section also presents a detailed data flow of the application.

In the end the authors argue that online social network security architectures are a powerful concept – yet to be discovered for future use by individuals and organizations alike, and that this work is a step in this direction.

The Chapter titled "*Computational Trust*" by B. Andersen et al. introduces the concept of trust for Internet of Everything (IoE). Further, the authors explain three main types of applicable trust that can be applied to trust in IT

- Trusting beliefs means a secure conviction that the other party has favorable attributes (such as benevolence, integrity, and competence), strong enough to create trusting intentions.
- Trusting intentions means a secure, committed willingness to depend upon, or to become vulnerable to, the other party in specific ways, strong enough to create trusting behaviors.
- Trusting behaviors means assured actions that demonstrate that one does in fact depend or rely upon the other party instead of on oneself or on controls. Trusting behavior is the action manifestation of willingness to depend.

Following sections presents high level computational trust engine and explain how trust (subjective probability of trust level) and trustworthiness (objective probability of trust level) can differ. Various computational trust algorithms have been discussed. The chapter is concluded with the examples where the notion of trust is explained in two well know internet services ie., PGP web of Trust and and X.509 certificates.

Where PGP trust model is a cumulative trust model. PGP uses a public-key ring, a structure that store public-key of other users known by the public key ring user. One or more certificates can be attached to the public key. PGP checks the public-key ring to see if the owner of the public-key is among the known public-key owners. If this is true, the SIGTRUST value is set to OWNERTRUST value. If not the SIGTRUST value is set to unknown user.

X.509 certificate contains the public key of the user and is signed with a key from a trusted CA (Certificate Authority) and is based on the use of public key cryptography. Regarding trust, it is obvious that there will be a common trust to the CA if all are using the same CA.

In the chapter *"Security in Internet of Things"* the authors Egon Kidmose and Jens Myrup Pedersen present a number of examples of problematic IoT devices and discuss the trade-offs which is often being made in designing the devices. The overall purpose of the chapter is to present motivation and suggestions for the design, implementation and deployment/configuration of systems that involved IoT devices. The authors discuss a number of IoT devices and use cases which are related to the IoT infra-structure and explain to which extend these devices are problematic seen from a security angle. Some of the discussions are about:

- IP Camera which in a study shows that there are significant security flaws when particularly allowing for connections over the Internet
- Internet Gateways as an important component in the IoT infra-structure where, amongst others, security not is set by default
- Smart Energy Meters which are the foundations of smart grids and already in for example in Ukraine has been subject to hacking and spearfishing attacks against central stakeholders of the electricity distribution
- Automotive IoT where there are several examples of the integration of systems in the smart car can be subject to hacking and eventually hostile take-over of the car
- Health IoT where the authors discuss the problems of hacking of smart devices which a specific health related purpose and the security weaknesses which is found in some of these devices

- Smart home appliances which often are set-up by the customers them-selves leaving a potential for not having sufficient knowledge on the stop of the systems relating to security and the lack of legislation to make sure that security is dealt with

All of the above-mentioned devices and use-cases are exemplified with real cases and situations in which security attacks and flaws have happened.

The authors use the above examples to present a number of security recommendations for manufacturers, developers and users which includes, amongst others, suggestions on practices such as: Incorporate security at the design phase, build on proven security practices, promote transparency across IoT, and use strong encryption for all communication. The chapter concludes that the development of the IoT infra-structure demands immediate action on deplorable security practices seen today.

Aske Hornbæk Knudsen, Jens Myrup Pedersen, Mikki Alexander Mousing Sørensen and Theis Dahl Villumsen have written a chapter on *"Security in the Industrial Internet of Things"*. The focus of their work is to set focus on industrial devices connected to the Internet such as robots and "smart production systems". The reason for focusing on industrial IoT is that many production facilities are critical for society and there can be economic as well as societal consequences if such systems are left inoperable of damaged by security attacks.

The chapter describes how a production line has been analyzed for security flaws. The work is part of a research project on Smart Production made at Aalborg University in 2016. The production system has been set-up without special consideration or care to security. The production line was not in a real production facility but resembled in many ways a real production line. The analyses made can therefore make grounds for demonstrating the con-sequences of setting up a production system without security measures. The production system was tested to see if the authors could reach the following three goals: add/change/delete orders; obtain performance information data as well as information on production costs without authentication; and cause severe damage to the system by deleting a large number of vital files.

The production line was tested via performance of different penetration tests. The method as divided into four steps: Describe the approach, Research, Scan and Find. The scan was done systematically by scanning all ports on the entire /15 subnet to get an overview of the machines and the production line. Afterwards a Wireshark was used to sniff traffic while an order of 10 products was being created in the production line.

The findings of the analyses were that the system was running on an open telnet connection that allowed access to the entire system as root. Due to the same default login credentials on all production devices with the user being the privileged root user, the authors gained full access to the system directly and could fulfill the above-mentioned goals of add, deleting, damage etc. of the system files.

The case study described in this chapter ends with a number of recommendations based on the observations. Amongst others, the authors provide specific recommendations to the set-up of the production line including changes in the infra-structure to firewalls and application of secure options such as SSH.

The case study demonstrates how it is possible to hack an automated production line with simple knowledge of the system. The paper concludes that the IoT integration of individual devices changes the role and usage of the individual device and therefore there is a need to upgrade old devices so that the system is not jeopardized by these.

Contributions – from Industry

"Modern & Resilient Cybersecurity" by Ole Kjedsen discusses the growing concern about Data & IT security in the industry. It is noted that the discussion have moved into both the board room of most companies, into governments but also grown to be a concern of many citizens/individuals. Hacking has become a global enterprise. This chapter evaluates the larger trends driving the cybersecurity domain and some of the emerging technologies and new paradigms that are likely to have the biggest influence on our digital lives and consequential the cybersecurity environment ending up in a list of recommendations for how to improve and further develop a modern, trustworthy & resilient cybersecurity strategy.

It is stated that important general trends can be summarized in these points

1. Cybersecurity threats and incidents are growing in numbers and sophistication.
2. Most attacks (including Ransomware) can still be avoided by following well-known advice and best practices.
3. Human error and/or simple fraudulent behavior (such as phishing) are behind most security breaches.
4. Countries and regions are very different in vulnerability profile and attack history/timing.

Noting this, it is stated that we are at a point, where we need to accept the reality of cybercrime – that most cyber criminals will never be caught and some will operate with near impunity leading to the strategy that instead of aiming to hurt/punish the criminal him/herself, it is better to deter and disrupt the business they run, and simply make it as hard as possible for them and their organization to succeed in their criminal endeavors. Further that regardless of how we approach the modern world of digital services – Protecting, Detecting and even add Deterence and Quick Response to our strategy, we will not end up in a zero-risk cyber environment. This calls for resilience and the agility needed to survive a security incident where the organization is prepared for the unavoidable system failure. To ensure the relevant resilience constant investments need to be made in education; Identification of best practices; Story telling around good and bad practices created delivery of modern tools; pervasive process guidance; and constantly nudging the user down the right path.

Technology trends are stated to add both complexity, new threats AND new capabilities in many dimensions of cybersecurity. Three of the most imminent and potent are Cloud Computing, Internet of Things and Artificial Intelligence. In this relation it noted that while these technologies do introduce a new attack surface and some new methods of attack for the cyber-criminal, most often they also introduce a very professional Protection as well as Detection and Respond mechanism/setup.

It is stated as a conclusion that the notion of an Eco-system of Cyber Security partners is pivotal for the success in creating the trust needed in internet driven solutions in the coming years. In such an environment, there is an all but obvious need for Public-Private Partnerships, collaboration between commercial competitors, collaboration between customers and vendors, vendors and law enforcement agencies, broad deployment of self-sustaining defense mechanisms and not least a fruitful continuation of the current discussion at the Governmental level regarding norms of behavior in cyberspace. The days where a vendor of software, network, hardware etc. could simply apply trustworthy methods to their code and/or device building and maintenance are gone.

In *"Building Secure Data Centres for Cloud based Services"*, Lars Kiergaard argues based on a Danish case study that Internet of Things (IoT) and cloud based services are key components in the emerging next generation business eco system. IoT systems will/ are collecting data from all aspects of daily and business life including communication, consumption, production, transportation etc. and cloud computing enables ubiquitous, convenient, on-demand storage, handling/manipulation and access to the data collected. It is noted that cloud based services offer significant advantages over traditional client-server based solutions not only in terms of scalability in the number of users that the service supports but also in the ability to offer the services across all client devices, i.e. smart phones, tablets, laptops and desktop pc's. Cloud-based services require that the frontend and backend software is hosted in data centres dimensioned for different purposes and different service levels.

Overall, data centres are defined and categorized into four different levels, or tiers, in accordance with an ANSI standard and representing different requirement s to guaranteed service levels and safety/security. Service level guarantees ranges from 99.671% for tier 1 to tier 4 where all components are redundant. Regarding security a tier 1 data centre would be sufficient for hosting a gaming service, whereas a public safety & emergency service providing situational awareness for civil preparedness require a tier 3 or a tier 4 data centre with disaster tolerance. It is noted that there is concern for privacy of the customer's data as cloud services potentially include combining user data from different sources potentially violating data protection rules as, e.g., the GDPR (General Data Protection Regulation) of the EU. This implies the implementation of additional security measures compared to traditional client-server systems, but apparently the standard for data centres does not address this issue.

The case study is describing the data centre of Teracom, a company providing the infrastructure for radio and TV broadcast in Denmark. The transition from analogue to digital TV distribution left a lot of physical space empty in the buildings, which housed the analogue TV transmitters and further some basic equipment for data centres, e.g., UPS (Uninterruptable Power Supply) and diesel generators. This was the foundation for Teracom to enter into the hosting or data centre business. The data centre solution has been designed as a Tier 3 data centre. The solution is based on dual-powered equipment, e.g. using UPS and diesel generators, and multiple uplinks to the data centre. This configuration offers a guaranteed service availability of 99.982%. The move from traditional broadcast to a provider of data centres has increased the requirements in terms of cyber security by many factors and as a new area also necessitated focus on privacy concerns. For this reason, Teracom is in the process of obtaining an ISO 27001 certification in IT security.

Kristoffer Rohde in *"Pervasive Governance – Understand and Secure Your Transaction Data & Content"* discusses the obligations included in the new EU General Data Protection Regulation (GDPR) that will require organizations to rethink how they store, manage and report on the information they own or process as part of doing business. Information is the lifeblood of any modern-day organization and information needs to be treated as a critical corporate asset and managed according to a well-defined information governance strategy. A number of challenges when designing and implementing an information governance strategy for today's complex business environment is outlined:

Exponential growth in digital content – "storing everything forever" is not cost-effective

A wide range of data & content types – many types of information must be retained to fully satisfy compliance requirements

Legacy and active applications – It is necessary to store content and data, which may span multiple obsolete or obsolescent applications; possibly introducing intelligent archival solutions

Increased pace of regulatory change and tougher sanctions – business and IT managers must respond to a wide range of rapidly changing rules and regulations

The chapter focuses on how organizations can understand and secure their unstructured content and transaction data. Different often seen approaches are discussed including the *Fragmented Approach* resulting from that many vendors have added an appearance of retention management to each of their products making it difficult for organizations to scale their retention management capabilities to across the enterprise; *the Classic Records Management Approach* that introduces a records management system with a broad set of system-enforced capabilities, but the time and effort required can make it impractical for wide-scale deployment; *the Keeping Legacy Systems Alive Approach* is expensive involving millions of Euro to protect and maintain old hardware, software and infrastructure and is usually unable to provide third-party access or reproduce entire transactions more than a couple of years ago.

Against these conventional approaches an ideal scenario a modular approach to records and retention management is presented around four main areas of capabilities: Enterprise Content Management, Core Retention Capability, Formal Records Management Capability and Archiving & Decommissioning. However, it is made clear that implementing The Ideal Scenario for managing transaction data and unstructured content assets in an end-to-end content compliance platform across the enterprise may not be a viable path for organizations in the short or medium term. It is stated that the best approach is to focus on protecting and securing information, while making it accessible and reusable based on policy and usage rights and permissions, regardless of its location. It is seen as an essential task to get an effective archiving & decommissioning program in place, organizations can reduce the cost of maintaining applications and static information, including structured and unstructured data, the contextual linkages between disparate data sources, and the software needed to provide access to that information. The program should be established based on a roadmap outlined in phases resulting in a model that is vendor independent, stable, and application neutral. The aim is an intelligent enterprise archive; where the decommissioning legacy systems reduce both reduce cost and risk, while improving the usability and compliance of historical and valuable corporate information moving forward. It is demonstrated that archiving following this vison not only eliminates the need for IT staff to maintain the old systems, but also means that information can be accessed in one place.

Fei Liu and Marcus Wong are the authors of the chapter on *"Challenges of Cyber Security and a Fundamental Way to Address Cyber Security"*. This chapter discusses the general challenges of cyber security seen from the service provider's viewpoint and uses this to point to the 3GPP security assurance approach. The chapter takes its basis in looking at what went wrong with the products form the telecommunication industry and identifies four areas that constitute a security problem:

- Functional designs – where products, such as graphical user interfaces and operating system designs have been developed with a focus on the functional design more than security and that this has caused vulnerabilities and threats to the users
- Proliferation of Internet – where most computers, industrial as well as personal, are online and that this Internet access has increased the potential for hacking since security not has been in focus access design
- Being a big target – where the authors discuss that big companies are targets for hackers etc. for financial purposes

- Quick to Market – where the authors describe that the process of rushing a product to market has some consequences in the lack of security solutions since these are often not done in a speeded up development cycle

The authors describe that there is a paradigm shift in cyber security and that this focuses on a combination between security by design principles, security by poof and security by assurance. They discuss that the market needs a security assurance that must be guaranteed and agreed on by the entire telecommunications community. This of course has a number of challenges such as for example regulatory and market place challenges.

The chapter ends by suggesting that the security challenges must be solved by establishing a security assurance process. This would set demands to vendors to follow strict operator and legal requirements as well as industry best practices build for the products to the highest degree of standards and the highest degree of security and integrity. The authors point to the 3GPP to secure a multi-layered security approach and secure the same kind of security around the world.

Concluding the paper, the authors state that security begins with a commitment coupled with a foundation of understanding of the threats, to define a security assurance process and to verify this.

References

- EU (2016): ec.europa.en/justice/data-protection
- Forbes (2016): https://www.forbes.com/sites/louiscolumbus/2016/11/27/roundup-of-internet-of-things-forecasts-and-market-estimates-2016/#2c644c1a292d
- Patel, K. K. and Patel, S. M. (2016): Internet of Things – IOT: Definition, Characteristics, Architecture, Enabling Technologies, Application & Future Challenges. International Journal of Engineering Science and Computing, Volume 6, No. 5, pp. 6122–6131.
- Security Intelligence (2016): https://securityintelligence.com/how-the-internet-of-things-iot-is-changing-the-cybersecurity-landscape/
- Westin, A. F. (1968): Privacy and Freedom. Washington and Lew Law Review, 25(1), 166.
- Ziegeldorf, J. H., Morchon, O. G. and Wehrle, K. (2013): Privacy in the Internet of Things: Threats and Challenges. Security and Communication Networks, 2013: 2728–2742.

1

An Introduction to Security Challenges in User-Facing Cryptographic Software

Greig Paul and James Irvine

Department of Electronic & Electrical Engineering,
University of Strathclyde, Glasgow, United Kingdom

1.1 Usability and Security

One of the key challenges in the development of secure software is the tradeoff between usability and security. Often, many of the rigorous requirements of a strong cryptographic implementation appear to be at odds with consumer requirements and desires. Non-technical users typically desire a straightforward user interface which does not require them to learn any special skills to use the application, yet also expect the application to offer them adequate protection [30]. There is, however, very little that an average user can do to ensure the security of the underlying technical implementation of security software they run, presenting a major challenge for users left unable to conveniently verify that the software works as expected. The intersection of the technical requirements for cryptography, and consumers' desires for usability introduces a number of opportunities for security weaknesses to emerge within the design of security software. A desire for convenience has been widely recognised as resulting in poor security practices, such as in the selection of passwords [31], of particular concern where user passwords are used for the generation of encryption keys for data.

In general, there are two main challenges when building a good security application. The first is to ensure that the underlying implementation is technically secure, with correct and proper implementations of cryptographic techniques, and resistant to attack. The second is to ensure that the software is suitably easy and intuitive for users to operate while doing so in a secure manner. If either of these challenges is not met, the resulting software will not offer practical security for users.

Within smartphone applications, other considerations are introduced; there is a very wide performance range among Android smartphones, which results in trade-offs being made to ensure good performance on all devices. Good performance is necessary to maintain user engagement. Monetisation models used for such applications also present an area of risk [23] with ad-supported applications the most common, the incentive structure for developers is weighted to reward building an application which is widely used and is able to serve a large number of adverts, rather than one which is less convenient but providing better security of user data. Ease of use has been highlighted previously to serve as a major factor in user retention [13], and the incentive structure of a free ad-supported app is such that retention of users will lead to more advert impressions. Previous work has considered the cost of free mobile apps, from the perspective of data usage [32]. While there exists considerable research on the privacy impact of ad supported applications, due to the use of advertising libraries [18, 22, 25], there does not appear to have been researching into the correct technical operation of such applications.

To highlight the risks for consumers, as well as the potential areas of concern when developing such software, a number of case-studies will be considered, using real encryption applications from the Google Play Store. The cryptographic operation of these applications will be discussed, based upon findings carried out using black-box analysis, by investigating the output of the application's encryption process. This will be used to identify a series of fundamental points which developers should follow when implementing an application making significant use of cryptography, in addition to ways in which a user may be able to determine if an application is protecting their personal data as they understand from the application description.

1.2 Background

One of the main pieces of background work on the usability of security and cryptography was carried out on the Pretty Good Privacy (PGP) email security product [29], highlighting the shortcomings of the software from the perspective of providing usable security. In their work, Whitten and Tygar undertook user testing to identify whether novice users could correctly complete a set of tasks without compromising the security of their messages during the test. Some of their main findings were that two-thirds of their test participants were unable to use PGP 5.0 to sign and encrypt a test email message within 90 minutes. One quarter of their test participants accidentally exposed the content of their test email message in plaintext, thinking that it had been encrypted when that was not the case. One of the main conclusions made was that for security to be usable by those who do not understand the underlying cryptography, the process of design needs to encompass more than the graphical interface alone.

Subsequent research by Sheng et al. [24] has highlighted that the problems identified within Whitten and Tygar's analysis remained the case, with no user in their test able to verify their keys, and many having issues with signing of the keys. Problems were also identified in the implementation of encryption, with none of the users able to identify when a message was being encrypted prior to sending [24].

This highlights two challenges – firstly to build a functional interface which facilitates users gaining the security properties they expect, without requiring an understanding of cryptography, and secondly to ensure that within this interface, the product itself is making proper use of cryptography to correctly encrypt user data.

Previous work has considered the usability of implementations of cryptography within end-user products, considering a range of encrypted USB flash drives and hard drives, as well as software-based solutions [26]. Nonetheless, such work has typically not focused on verifying whether or not the implementation of cryptography within the application is correct and suitable.

1.3 Practical Cryptographic Implementation

When creating software which implements cryptographic algorithms, there are a number of important considerations. For user-facing software, one of the primary concerns is how the key is stored or entered. Symmetric cryptographic keys are typically 128 or 256 bits for the AES cipher. These keys must be generated using a strong random number source, and be unpredictable to anyone not already knowing them (i.e. keys should not be incremented sequentially or otherwise correlated). One common design pattern is to allow the user to enter a password or passphrase, and use a key derivation function (KDF). This KDF should take a string such as a password or a passphrase as the input, and derive a uniform, unpredictable output, which is suitable for use as a cryptographic key. Examples of such KDFs include PBKDF#2, scrypt [21] and similar. By deriving a key from the user's entered password, and ensuring it is properly discarded from memory after use, it is not possible to re-derive this key without the user's credential being entered. Figure 1.1 illustrates the operation of a KDF to derive a key from a user's password, passphrase or another credential.

Figure 1.1　Flow chart indicating the operation of a key derivation function, to derive a cryptographic key from a user-supplied password or passphrase.

This approach is commonly seen in desktop file encryption software, where the user is directed to enter a strong password or passphrase, which is then used to derive a key. One limitation of this approach is that a user must either enter their passphrase for each cryptographic operation or allow it to be cached by the software carrying out the operations. Caching the key and leaving it in memory puts the key at risk, as it will be present even when not in use, leaving it exposed to other software running on the computer. By minimising the time the key is exposed, the risk of malicious software being able to compromise the key from memory is decreased. Another important limitation of this approach is that it is not possible to allow a user to change their password without decrypting and re-encrypting any data protected using that key – it must first be decrypted to a plaintext, then a new key derived from the new passphrase, then encrypted using the new key to produce a new ciphertext. This process may be very resource intensive for a large number of files, or large quantity of data. Figure 1.2 shows how an encryption tool could be formed around a password-derived key.

An alternative approach which permits the user to freely alter their password without re-encrypting all data is to use two separate keys; one being derived from the user's passphrase through a KDF, as previously, in addition to a *master key*, which is randomly generated from a secure random

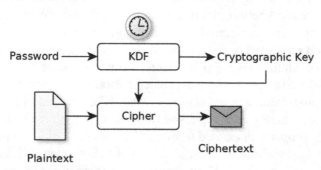

Figure 1.2　Flow chart indicating the operation of a simple encryption utility, using a password-derived key to encrypt data with a cipher.

number source. The randomly generated key is then encrypted using the key-encryption key (KEK), which is the key derived from the user's passphrase. The random key is then used to encrypt actual data or files. This approach is commonly used in filesystem-level encryption, for example in tools such as Truecrypt and Veracrypt. Figure 1.3 shows how a master key can be introduced alongside a password-derived key to protect user data without requiring re-encryption when the user's credential is changed.

To permit a user to change their password under this approach, only the master key need be re-encrypted; it is first decrypted using the previous credential-derived key, and then re-encrypted using the new credential-derived key. The master key is not changed in this process, and therefore files encrypted with it need not be re-encrypted. A weakness of this approach is that the same key is used to decrypt each file, and therefore it is critical that a suitable cipher-mode is used, which is properly initialised using a random initialisation parameter for each file. This is necessary to prevent attackers from comparing ciphertexts. For example, for the AES cipher in counter mode, $C_1 \oplus C_2 = P_1 \oplus P_2$ for any two ciphertexts C_1 and C_2, and two plaintexts P_1 and P_2, where the same key is used for both encryption operations. This therefore leaks information between messages, and allows an attacker knowing the plaintext of a message and its given ciphertext to utilise that to decrypt any other ciphertext which was encrypted using the same key.

This attack may be mitigated through the use of an initialised cipher mode, such as CBC (cipher block chaining), where an initialisation vector or other

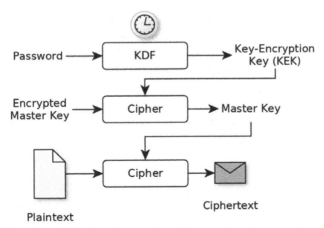

Figure 1.3 Flow chart indicating the operation of a more complex encryption utility, using a password-derived key to protect a master key, which is then used to encrypt data with a cipher.

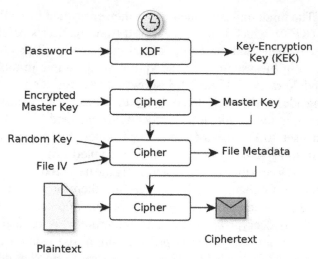

Figure 1.4   Flow chart indicating the operation of a more complex encryption utility, using a password-derived key to protect a master key, which is then used to encrypt a per-file metadata block, itself incorporating a file encryption key and IV, which are used to protect the file itself.

per-use parameter is set. The initialisation vector (IV) need not be kept secret, and usually need not be random, provided it is not re-used. A common approach is to prepend the IV to the ciphertext output, since the IV length is fixed and based upon the block length of the cipher in use.

An alternative approach to mitigate the risk of key re-use is to use a unique key for each file or record which is encrypted. In this case, a set of metadata is created for each ciphertext, with this metadata incorporating the key and IV used for each file. The user's credential-derived key is used to decrypt the *master key*, which is then used to decrypt the metadata record for a given file or record, thus obtaining the key and IV for that file. This approach is frequently used in file or record-based encryption, where each file must be independently accessed, and risks of potential key-reuse are mitigated. Figure 1.4 shows an approach to use a per-file key and IV, for a scenario where each file is separately keyed, and files do not require re-encryption to permit a user to change their password.

1.4 Analysis of a Selection of Android Encryption Apps

A selection of Android encryption apps has been analysed. A more detailed explanation of this work and the methodology used is available within [19].

Nine encryption applications were investigated, with 5 of them published on the Google Play Store by "top developers". These applications were mostly designed for the encryption of photos and videos. One application supported files of any type, and one was a password manager, produced by one of the developers which also produced a photo and video encryption application. The applications investigated were relatively popular, with millions of users in many cases – considering the Google Play Store's upper bound on the number of installations of the application, the applications analysed could potentially be in use by as many as 117 million users.

Each application was used to encrypt a file of a suitable type depending on the application in question. The resulting encrypted file was then retrieved from the shared storage of the device. The original plaintext file was then compared with the retrieved ciphertext. This replicated an attack consistent with another piece of software on the smartphone exfiltrating files which had been encrypted by a user. Therefore, if these files were able to be decrypted, other software on the smartphone would be able to overcome the confidentiality offered by the encryption application in question.

To verify the integrity of recovered files, where it was found to be possible to obtain the original data from analysis of the ciphertext alone, the original plaintext was compared bytewise against the recovered plaintext, to ensure the recovered data was identical.

1.4.1 Main Findings

Of the 7 applications analysed which were specifically designed and marketed as being for the encryption of photographs or videos, all 7 were found to leave the vast majority of the file intact, with only a small header of the file differing between the plaintext and ciphertext. In one case, each of the first 100 bytes of the file header had been flipped (i.e. subtracted from the hexadecimal value 0xFF). In another case, the first 10 bytes of a JPEG file header was removed from the file, and replaced with zero byte values. The remainder of the file was untouched, and the bytes taken from the file header were stored in a SQlite database, itself located in the shared storage area, thus permitting any application to easily retrieve the original first ten bytes.

Although none of the applications for the photograph and video encryption was found to protect more than the header of the file, some did make use of encryption algorithms rather than mere obfuscation for these headers. These were therefore investigated, to ascertain whether or not the application would offer sufficient security if the application was modified to encrypt the full file.

A number of significant issues were however found in the implementation of this encryption.

To investigate these applications, two photos were input to the application, and thus two ciphertexts were retrieved. The input files will contain similar initial contents, since the file headers of image files include significant replication. The headers of the two plaintexts were then compared, along with the corresponding headers of the two ciphertexts. It was immediately apparent that the two ciphertext outputs for different files were almost identical, with only occasional bytes differing between both header outputs. This indicated that it was most likely a stream cipher mode of operation was in use, since a block-based cipher would produce a significantly different output block (of 16 bytes, for the case of AES) for a slightly changed input block.

The differing bytes were found to be at the same offsets within the headers; if the 20th byte of the two plaintexts differed, the 20th byte of the ciphertexts would also differ. It was then identified that $C_1 \oplus C_2 = P_1 \oplus P_2$ for ciphertexts C_1 and C_2, and plaintexts P_1 and P_2. This meant that the cipher in use was operating with the same key and initialisation parameters between subsequent operations on different files. Further analysis using the above techniques, comparing ciphertext output across different devices revealed that the same key was used on all devices running this application, and that therefore the cipher key and initialisation parameters were likely hard-coded within the application.

This meant that any third party was able to decrypt the header of files encrypted using this application, as a result of the fixed key and initialisation parameters.

The password manager application investigated was published by developer of the photo and video encryption applications which used fixed encryption keys and parameters. By carrying out similar analysis of the database entries within the password manager, it was identified that a static key and initialisation parameter was also in use within that application; common prefixes in ciphertexts were found to correspond to common prefixes in plaintexts. By repeating the above process, it was possible to recover the plaintexts of passwords encrypted on one smartphone, simply by using the application on another device to create a known plaintext and ciphertext combination.

In all of these cases, the user's selected credential was a PIN, entered within the application. The PIN was not used to derive an encryption key in any of the applications, and was used only to pose an aesthetic measure of reassurance that files were protected; it was not necessary to have knowledge of the user's PIN to recover plaintexts from any of the apps analysed.

For the application supporting the encryption of files of any type, it was identified to be implementing a form of monoalphabetic substitution cipher, due to the one-for-one propagation of patterns within the input plaintext to the output ciphertext. A header was prepended to the ciphertext, and analysis of this revealed that there was a 256-byte block, within which each byte value from 0x00 to 0xFF appeared once. This block served as a lookup table for mapping plaintext and ciphertext bytes. The application was therefore merely implementing a form of obfuscation, rather than encryption of the file. In addition, the unsalted MD5 of the user's PIN was exposed within the header of each ciphertext, allowing any software with access to a ciphertext to identify a user's plaintext PIN efficiently through the use of a rainbow table or exhaustive search, since MD5 does not offer any security against brute force.

1.5 Priorities to Improve upon Existing Applications

To improve upon the weaknesses identified within existing applications, several improvements could be made. The top priority would be to ensure that the full contents of a file is encrypted, rather than merely the headers, and to ensure that a standard, strong symmetric encryption algorithm is used for this process, rather than obfuscation or any custom ciphers.

The next priority is to ensure that the user's credentials are directly used to derive an encryption key through a suitable key derivation function, and that the user's password or passphrase contains sufficient entropy to be used safely as a key – a 4-digit PIN does not, as will be explored in Section 1.6.1. Although it is possible to bind a cryptographic key to a device-specific security engine, through the Android keystore interface, this would make encrypted files unusable without access to the device used originally for encryption, and would thus make encrypted files non-portable, and unrecoverable with another access to another device and the original credential. This raises a serious concern of data loss if the original device was no longer accessible.

After ensuring that the user's credentials contain sufficient entropy to be used as a key, the next priority would be to ensure proper cipher mode usage; with some of the applications re-using initialisation vectors in stream cipher modes such as AES-CTR (counter mode), it is necessary to ensure that a suitable cipher mode is used to prevent attacks between pairs of plaintexts and ciphertexts.

1.6 Implementation Considerations

Having identified a suitable cryptographic approach in Section 1.3, and explored the limitations and weaknesses of a number of current encryption apps, leading towards some priority issues to resolve in Section 1.5, it is necessary to ensure that the implementation used offers the necessary security properties to protect data adequately. By breaking the process down into steps, the requirements at each step can be identified. Note that at this point, it is assumed that well-reviewed standardised ciphers, which have been subject to significant security, have been used. Examples of appropriate cipher constructs would include AES, for example.

1.6.1 Key Derivation Stage

At the step of key derivation from a user's password, two key assumptions are made. The first is that the user's password or passphrase offers sufficient entropy for the derivation of an unpredictable key. The performance of an exhaustive attack against the output of a KDF is linear with the number of candidate inputs the function is evaluated for. It is also possible to carry out multiple operations in parallel, resulting in a potentially significant increase in attack throughput. Therefore, even though a good KDF will have a significant time penalty in the computation of the output, it is still necessary to ensure that the input comes from a suitably large range of potential inputs, such that they cannot be feasibly evaluated exhaustively. By way of example, even with the use of a very slow KDF, tuned to take 5 seconds for an output to be computed, a 4-digit PIN could be exhaustively identified in around 14 hours, using a single processor. This could be reduced to around 13 minutes if 64 processor cores were used in parallel. With the advent of cloud computing, and the ability to rent highly powerful computing resources for a short period of time, attacks such as this are highly practical.

Therefore, it is essential to ensure that the user selects a suitably strong passphrase, which is sufficiently unpredictable to make such attacks on the master password infeasible. This is potentially a challenge for usability, since forced password length and complexity are particularly undesirable on smart-phone keyboards and other portable devices, where it is difficult or inconvenient to access numerical and symbol keyboards [6].

At this stage, it is also important to ensure that the KDF uses a salt which is highly likely to be unique. While it can never be guaranteed that a salt is globally unique, selecting a 128-bit random salt should offer sufficient collision resistance to suffice. The salt is necessary to prevent a pre-computed table of hashes being used to rapidly evaluate commonly-chosen user passwords, without requiring the process of slowly computing the KDF function output for each individual attack. Such pre-computed tables of hashes are typically called *rainbow tables*, and are commonly available for a variety of unsalted password hashing schemes.

1.6.2 Master Key Generation and Use

When a master key is used, and encrypted by a key derived from the user's chosen credential, the security of the generation of this master key is critical. If this key were able to be obtained through a side channel means, it would bypass the need for knowledge of the user's credential to gain access to protected files. If side channel attacks on the computer's memory are discounted, one of the main ways for this key to be compromised would be through the use of a poor quality entropy source for the generation of the cryptographic key.

The process of gaining high quality random numbers from computers is a well-studied topic, although one which is commonly misunderstood, due to a number of long-running misconceptions surrounding the Unix /dev/random and /dev/urandom devices [11].

On mobile devices, the security of random number generation is something which should certainly be considered; if a mobile device has been compromised by malicious software, it is possible for the output of the random number generator, running within the operating system's kernel to be compromised from regular software or apps running on the device [20], resulting in the production of predictable or known output when random numbers are requested. Previous work has investigated the security of random number generation on the Android platform, both at kernel level [14] and at application level [10] in Bitcoin and other applications. Indeed, Bitcoin wallet applications, which themselves rely heavily on the ability to generate strong random numbers for private keys, have previously faced other issues due to implementational errors – bugs in the implementation of SecureRandom were found to result in poor quality random numbers being made available to applications [9].

The legacy SHA1PRNG pseudo-random number generator, which has recently been deprecated by Google [7], also presents a potential hazard for developers, since it is commonly used for the generation of random numbers for cryptographic keys. As of January 2017, there are in excess of 20,000 results found when searching GitHub, a commonly used source code sharing website, for the string SHA1PRNG, which indicates that this is still relatively widely used. Previous work has highlighted biases in the output of this pseudo-random number generator algorithm [28], and developers should follow the guidance from Google to correctly generate random numbers more securely [7].

Therefore, it is important to ensure that where a master key is generated randomly, the random number generation APIs are used correctly. For the Android platform, Google makes available advice on the correct use in their security blog [16].

1.6.3 Cipher Use and Initialisation

Where a master key is encrypted by a key derived from a user credential, the correct use and initialisation of the relevant cipher are essential, in order to protect the security of the master key. The same is also true when the underlying data itself is encrypted for storage, using either a master key or a per-file metadata key. To prevent attacks such as those explored in Section 1.4.1, it is necessary to ensure that all cryptography is used correctly.

The first requirement is that a suitable cipher must be selected. This should be widely adopted, a publicly available standard (per Kerckhoffs' assumption [15]), and have undergone peer review from the cryptography community. One example of a suitable and widely-used cipher is AES. AES operates as a block cipher, with blocks of 16 bytes. Therefore, if a plaintext is longer than 16 bytes, it will be necessary for the cipher to be used on multiple blocks. The most naive approach to this, which introduces a number of vulnerabilities, is the use of Encrypted Code Book mode (ECB). In ECB mode, each block is independently encrypted using the same key. The result is that patterns at block level will be transparent within the ciphertext, since the same input block of 16 bytes will always produce the same ciphertext output of 16 bytes for any given key. To mitigate this, a suitable cipher mode must be chosen, which ensures that a repeated cipher input block will produce an output block which is indistinguishable from the previous block, as well as unrelated blocks. Figure 1.5 illustrates this, for a scenario where the same input to the cipher results in an identifiable repeated output block.

By chaining the output of the previous block of the cipher as an input to the next block, it is possible to ensure that different blocks will encrypt to different ciphertexts, thus preventing the identification of patterns within the plaintext from analysis of the ciphertext. This is commonly referred to as Cipher Block Chaining (CBC) mode. A special case is the first block, which

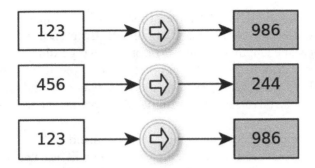

Figure 1.5 Demonstration of the risk of an uninitialised cipher producing identifiable ciphertext output blocks, revealing information about repetition patterns within the plaintext.

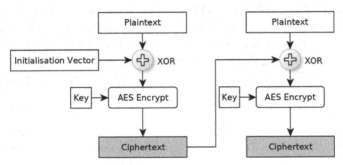

Figure 1.6 Illustration of cipher block chaining being used to encrypt a second ciphertext block based upon the previous block output, and an initialisation vector used for the first block.

takes an arbitrary initialisation vector (IV) as the input for the previous block's output, on account of there being no previous block to refer to.

The initialisation vector (IV) is not required to be kept secret, but must remain associated with the ciphertext, in order to ensure that the cipher can be correctly re-initialised before decryption. Therefore, the IV is often stored alongside the ciphertext, or within the corresponding metadata for the file to which it pertains, as discussed in Section 1.3.

1.6.4 Indistinguishability and Resistance to Malleability

Desirable properties of a high quality cipher and cryptographic implementation can be formulated intuitively, based upon properties which minimise the information available to an adversary or other party with access to ciphertexts. These properties have also been more formally defined and related in the literature [1]. Firstly, it is a desirable property for two ciphertexts to be opaque to a third party; if an adversary can tell the difference between two ciphertexts of the same length, this implies information regarding the contents is being revealed by the cipher. This property is referred to as ciphertext indistinguishability [8], requiring that for any two ciphertexts and plaintexts, it is not possible for an adversary to determine which plaintext corresponds to each output.

When considering these attacks, two oracles are considered. An oracle is a hypothetical black box function. An encryption oracle will encrypt any given plaintext, returning the ciphertext. A decryption oracle will decrypt any given ciphertext, returning the plaintext. These oracles can be used to define the scenarios under which a cipher will preserve security of data under certain attacks.

Ciphertext indistinguishability can be considered under multiple attack scenarios, presenting formally defined security properties. To seek indistinguishability under chosen plaintext attack (IND-CPA), an adversary should be able to select a plaintext, P, and be provided two ciphertexts C_1 and C_2 and be unable to determine whether C_1 or C_1 corresponds to the encrypted form of plaintext P.

Indistinguishability can also be considered under the chosen ciphertext attack models, where the attacker is permitted to generate as many ciphertexts as desired, with full access to both encryption and decryption oracles. The attacker then selects two different plaintexts, and must distinguish between the two corresponding ciphertexts. After selection of plaintexts, the decryption oracle is unavailable to the attacker. indistinguishability under a chosen ciphertext attack in this model is referred to as IND-CCA1 security.

An extended version of this attack scenario can be considered whereby adaptive attacks are still permitted; the decryption oracle is not removed from the attacker, who may carry out any operation other than decryption of the resulting ciphertexts. This attack scenario is referred to as IND-CCA2 security.

In terms of the security properties offered by each variant of indistinguishability, the strongest security is provided by resistance to adaptive chosen ciphertext attacks (IND-CCA2). This encompasses the properties of IND-CCA1 and IND-CPA. Likewise, IND-CCA1 encompasses the property of IND-CPA, since a chosen ciphertext attack also contains the plaintext selection step of a chosen plaintext attack.

Inkeeping with ciphertext indistinguishability, another key desirable property of a cipher or cryptographic implementation is its resistance to malleability. Malleability refers to the scenario where an attacker is able to manipulate a ciphertext C_1, corresponding to a plaintext P_1, such that upon decryption, the modified ciphertext C_1' can be decrypted to a function of the original plaintext, P_1', without requiring knowledge of the original plaintext.

Ciphertext malleability [5] presents a risk where a sensitive operation is carried out using an encrypted message, such that an adversary being able to carry out a small modification to a message may be able to influence the output in a predictable way. If the integrity of a ciphertext is not suitably authenticated, as discussed in Section 1.6.5, a malleable cipher may permit an adversary to make predictable modifications to the output of the cipher.

Like with indistinguishability, the property of resistance to malleability can be considered under the same three scenarios [3]; with a chosen plaintext, with a chosen ciphertext, and in an adaptive chosen ciphertext scenario. To ensure the integrity of data stored using a cryptographic implementation, it is desirable that the cipher not be malleable under any of the above scenarios.

1.6.5 Authentication of Ciphertexts

One area worth consideration for new designs of security software is the authentication of ciphertexts. With the rise of cloud storage and other systems whereby users may hold their data externally, some risks are introduced whereby a malicious attacker may attempt to manipulate the underlying plaintext through alterations made to the ciphertext. This could be carried out across a network, or by an entity providing storage to the victim. Alternatively, corruption of the storage medium may be the cause of modifications to the ciphertext when held at-rest. To avoid attacks such as this, the ciphertext output should be authenticated, to allow a user to determine that a ciphertext has been modified, and thus avoid relying upon the data contained within it.

Ciphertext authentication refers to the process of ensuring that they have not been tampered with or otherwise modified between being produced and being used. To ensure that a ciphertext cannot be deliberately modified by a third party, attempting to trick a user into decrypting a modified ciphertext, it is necessary to ensure that ciphertext authentication offers security beyond merely a checksum or similar cryptographic hash, since such a checksum could be modified by the attacker to be valid. It is therefore necessary for ciphertext authentication to be based upon a secret, known only by a legitimate party to the encryption. Typically, for file storage encryption, this authorised party would be the user. Therefore, the ciphertext should be authenticated by a means which prevents anyone else from generating valid authentication data for a modified ciphertext. Authentication therefore, provides assurance as to the legitimacy and origin of a ciphertext, and allows a user to verify that it has not been tampered with since it was created.

As discussed in Section 1.6.4, a determined attacker may be able to make a modification to a ciphertext, which will result in a predictable modification to the underlying plaintext message. This could be significant in scenarios where a user is relying on the encryption to prevent their data being modified or tampered with. For example, financial accounts or a to-do list could be tampered with to potentially alter the storage data. This happens because during decryption in CBC mode, any changes made to the ciphertext of block N will propagate directly to be XOR'd with the output of the cipher for block $N + 1$.

Figure 1.7 illustrates this by showing the operation of decryption in CBC mode.

There are four main approaches to ciphertext authentication. Three of these are based upon the use of a keyed authentication function. An example of this is a symmetric Message Authentication Code (MAC) or HMAC, which is a MAC derived from a cryptographic hashing function. Alternatively, authenticated encryption modes exist (including, but not limited to GCM mode), which can be used instead of cipher block chaining (CBC). These authenticated modes can be used to produce a ciphertext output which is verifiable for authenticity, in addition to offering confidentiality. A modified ciphertext will be detected,

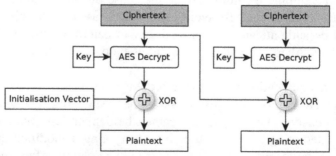

Figure 1.7   Illustration of cipher block chaining being used to decrypt two ciphertext blocks. Note that the output of the second cipher block (i.e. the plaintext) is derived from the cipher block output XOR'd with the previous ciphertext block, showing that arbitrary modifications can be made to the underlying plaintext block by manipulating the ciphertext.

and the decryption process will alert the user to the ciphertext having been tampered with, or refuse to proceed with decryption.

For ciphertext authentication based upon a keyed authenticator, such as a MAC, the three available approaches pertain to the order of the encryption and authentication operations. Since the MAC can be considered as a function which produces an output based upon any arbitrary input, it is possible to alter the order of the MAC and encryption operations. This leads to three constructs; Encrypt-then-MAC (ETM), MAC-then-Encrypt (MTE), and Encrypt-and-MAC (EAM).

The ETM construct is the preferred approach [3], whereby authentication is carried out on the output of the encryption cipher, meaning that the ciphertext is authenticated. One key advantage of this approach is that the authenticity of a ciphertext can be validated prior to any decryption operation, reducing risks of side channel attacks or implementation weaknesses being exploited, since only valid ciphertexts should be presented to the cipher for decryption.

The MAC-then-Encrypt (MTE) construct is another option available, where the plaintext is authenticated, and the resulting authenticator output placed into the cipher for encryption. A disadvantage of this approach is that it is not possible to validate a ciphertext until after it has been decrypted, since the authenticator itself is encrypted and cannot be used for

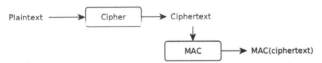

Figure 1.8 Illustration of the Encrypt-then-MAC (ETM) construct, where the plaintext is first encrypted, resulting in a ciphertext. This ciphertext then acts as the input to the MAC function, producing an authenticator output for the ciphertext itself.

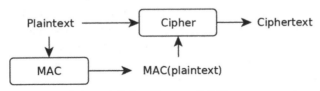

Figure 1.9 Illustration of the MAC-then-Encrypt (MTE) construct, where the incoming plaintext is passed through the MAC authenticator function initially. The output of this function, the authenticator, is then appended to the plaintext, and the result is then encrypted by the cipher, producing a ciphertext containing both the plaintext and MAC of the plaintext.

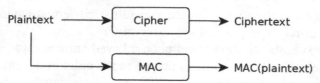

Figure 1.10 Illustration of the Encrypt-and-MAC (EAM) construct, where the encryption and authentication processes are carried out independently. The resulting MAC encompasses the plaintext, such that the ciphertext must first be decrypted, and then the computed plaintext authenticated against the authenticator function.

validation until following decryption of both the message and authenticator. This raises potential for attacks where a user is tricked into decrypting an untrusted ciphertext, which can lead to vulnerability to chosen ciphertext attacks [2, 17].

Finally, the Encrypt-and-MAC (EAM) construct operates by carrying out authentication of the plaintext. An advantage of this approach is that both the ciphertext and authenticator values can be computed at the same time; ETM requires encryption to take place before a MAC is calculated, and MTE requires the MAC to be computed before it is included in the ciphertext. A disadvantage of this approach is that during decryption it requires the decryption process to have taken place before the message can be authenticated. Therefore, this also gives rise to potential vulnerability to chosen ciphertext attacks [2, 17].

1.6.6 Padding Attacks

Where arbitrary-length messages need to be supported for encryption and decryption, it is necessary to consider how shorter blocks should be handled by a block cipher. Where a message length is not a multiple of the cipher block length, it is necessary to pad the final block's plaintext, in order to ensure that the block's length is equal to the cipher's block length. One naive approach to padding would be to append zeroes (or another fixed value of a byte) to the end of the block, but this can lead to ambiguity where a message may end in zeroes, which would be assumed to be padding and thus removed from the plaintext being returned. Therefore, it is necessary to have a more advanced padding scheme.

Padding schemes can be used to carry out attacks against ciphers however, and therefore correct plaintext padding is an important part of the overall security of a system. The PKCS#7 padding scheme is a widely used one [12], whereby N padding bytes are appended to a message, and each padding byte takes the value of N. For the case of no padding being required, a final block is added to the plaintext, entirely compromised of padding. This scheme of padding prevents ambiguity over the boundary between data and padding in a message. Message padding can however be a source of vulnerabilities.

If a block cipher is used in a mode such as CBC, it is possible to utilise the handling of padding to decrypt an arbitrary message, if it is possible to identify when incorrect padding resulted in a failure to decrypt. Whether or not padding succeeded can provide a small quantity of information, and this can iteratively be used to decrypt an arbitrary message. This type of attack is typically referred to as a padding oracle attack [27], since all it requires is a ciphertext, and a function or facility to determine whether a given decrypted message was correctly padded.

To carry out a padding oracle attack, a random block is generated, and a given block of ciphertext data is appended to it. A query is made to the padding oracle to determine if the padding on this message is correct, which is highly unlikely for an arbitrary ciphertext block. Until the padding is deemed correct by the padding oracle, the attacker needs to attempt to create correct padding by manipulating the last byte of the target ciphertext block to have a value of 0x01. To do this, the last byte of the random data block is modified; due to the chaining action of CBC as discussed previously, a modification of the last byte of block N will propagate to block $N + 1$. By attempting all 255 possible values for this byte, the attacker will eventually be informed the message has valid padding.

At this point, the correct length of the message must be identified, by sequentially modifying each byte of the initial random block. These modifications will propagate to respective blocks in the target ciphertext block. By carrying this out from left to right (i.e. from first to the last byte of the block), it will become apparent how many bytes of data are present in the message; as soon as modifying a byte causes padding to fail validation, the byte modified was the first byte of padding. Typically, there will only be 1 byte of padding, since the probability of happening across a plaintext ending in another valid padding sequence is exponentially lower for increased numbers of padding bytes, although this process will identify the padding length irrespective of this.

At this point, the attacker has decrypted all of the padding bytes' values, since an attacker knows that for each padding byte in the plaintext, PT_i, it can be XOR'd with the corresponding random byte to result in a value of 0x01 (the number of padding bytes identified), such that $PT_i \oplus R_i = pad$, where r_i is the corresponding byte from the randomly generated first block, and pad is the number of padding bytes identified for the message, typically 1. Therefore, the plaintext byte can be recovered: $PT_i = pad \oplus R_i$.

To decrypt the remaining bytes of the ciphertext, the attacker can XOR the final byte with the number of padding bytes present (pad, usually 1), and then XOR with $pad + 1$. This will set the last padding byte to the correct next, and permit the attacker to attempt each $pad + 1$ byte value, until the padding validates. The process is repeated for the full length of the ciphertext, thus recovering the full plaintext from a given message, by virtue of knowing whether a message was correctly padded.

1.7 Discussion

Fundamentally, the process of determining whether or not a given application correctly implements cryptography, such that users can be reassured their data is protected, is not an easy task. Short of carrying out an analysis of the ciphertext outputs of each application, it is difficult for users to gain any information as to the quality of implementation. It is clear from the results identified in the analysis discussed above that many applications currently in use today are not making even the most basic efforts to protect the confidentiality of user data, and that many apps labelled as implementing encryption were not doing so correctly. Specifically, each of the photo and video encryption apps were merely modifying the header of the file, while leaving the vast majority of the file (typically all of the file after the first 2 or 8 kilobytes) untouched. Those which were not merely obfuscating the file header were using static keys and initialisation vectors, rendering their encryption effectively worthless beyond obfuscation.

There is no clear way for users to determine whether or not a given encryption app offers suitable protection, given that some of the apps analysed were by Google Play Top Developers, and many had very high numbers of downloads, some in excess of 10 million. Looking for applications which adhere to Kerckhoffs' principle, and make their source code available for analysis would be one potential way to look for better applications which are willing to open their implementation to review and scrutiny, but at present this appears to be relatively uncommon, with none of the 9 apps analysed doing so.

One way in which there is potential for the situation to be improved would be if cryptographic libraries and APIs focused on providing higher level primitives for application developers to consume without expert cryptographic knowledge, thus attempting to minimise the potential for developers to make mistakes. This is the approach taken by much of the work of Bernstein [4], which focuses on building strong primitives with simple and accessible APIs for developers to use. By way of example, by making the default usage of an AES cipher require a randomly generated key and IV, developers' attention may be drawn to the operation of the cipher, and the need to use a 128 or 256-bit key, rather than attempt to use a 4-digit PIN. Greater awareness among developers of security applications is undoubtedly necessary, although without a cryptographic background, it is likely that there will continue to be vulnerabilities in applications, particularly around some of the more complex areas of cryptographic implementations. Examples such as padding attacks show how secure cryptographic algorithm implementation design is anything but intuitive, and requires the correct usage and understanding of cipher modes, and of key storage and generation. More open source implementations of usable cryptography-enabled applications may serve as examples for cryptographically-aware developers to refer to and utilise, which may help to reduce the number of weak implementations being made available.

1.8 Conclusions

Based upon analysis of a number of Android encryption applications, it is clear that at present there are several applications available today, offering to encrypt user data for confidentiality, which are failing to do so adequately. This is a problem for users, who wish to find a usable and simple encryption application, and are unable to verify the correct operation of the application using technical means, such as black-box testing and comparison of ciphertexts and plaintexts. A number of recommendations have been made for developers of encryption applications, based upon the issues identified within those investigated; these ranged from failing to use encryption at all, and carrying out merely basic obfuscation of file headers, through to using static, hard-coded keys and initialisation vectors for ciphers. In almost all cases, applications only modified the file headers, leaving the vast majority of the file data intact in plaintext form. This poses a significant risk for users who believe they are encrypting their data. In addition, improper use of keys meant that the PIN that a user was asked to provide was not used as an input to the encryption process, and rather as an access control layer to the application itself, with static encryption or obfuscation operating underneath, with a hard-coded process or key.

A number of recommendations have been made to improve the security of applications such as these being developed. These include ensuring good quality random number sources for all key generation processes, and using a master key for file encryption in addition to a user credential-derived key for master key protection. The importance of correct cipher use and implementation was also highlighted, along with the need to ensure a suitable mode of operation for the cipher is used, which prevents patterns in the plaintext from being translated through to patterns in the ciphertext. The need for ciphertext authentication to prevent tampering with plaintexts by sophisticated attackers has also been highlighted, along with an explanation of some of the more complex risks, such as padding oracle attacks, where a padding verification process can be used to recover plaintext data from a ciphertext. These recommendations serve to highlight the extent to which correct cryptographic implementation is not a simple task, and that application developers using cryptography should seek ensure they follow best practice. To perhaps reduce occurrences of implementation issues such as these, a focus on high level, verified, application-focused cryptographic libraries would be beneficial, to reduce the potential for application developers to compromise the security of their application through the need for complex cryptographic implementations in their application itself.

References

[1] Mihir Bellare, Anand Desai, David Pointcheval, and Phillip Rogaway. Relations among notions of security for public-key encryption schemes. In *Annual International Cryptology Conference*, pp. 26–45. Springer, 1998.

[2] Mihir Bellare, Tadayoshi Kohno, and Chanathip Namprempre. Breaking and provably repairing the SSH authenticated encryption scheme: A case study of the encode-then-encrypt-and-mac paradigm. *ACM Transactions on Information and System Security (TISSEC)*, 7(2):206–241, 2004.

[3] Mihir Bellare and Chanathip Namprempre. Authenticated encryption: Relations among notions and analysis of the generic composition paradigm. In *International Conference on the Theory and Application of Cryptology and Information Security*, pp. 531–545. Springer, 2000.

[4] Daniel J Bernstein. Cryptography in NaCl. *Networking and Cryptography Library*, 2009.

[5] Danny Dolev, Cynthia Dwork, and Moni Naor. Nonmalleable cryptography. *SIAM Review*, 45(4):727–784, 2003.

[6] Chris Foxx. How to pick the perfect password, September 2015. http://www.bbc.co.uk/news/technology-34221843.

[7] Sergio Giro. Security crypto provider deprecated in Android N, June 2016. https://android-developers.googleblog.com/2016/06/security-crypto-provider-deprecated-in.html

[8] Shafi Goldwasser and Silvio Micali. Probabilistic encryption. *Journal of Computer and System Sciences*, 28(2):270–299, 1984.

[9] Dan Goodin. Google confirms critical Android crypto flaw used in $5,700 Bitcoin heist, August 2013. http://arstechnica.com/security/2013/08/google-confirms-critical-android-crypto-flaw-used-in-5700-bitcoin-heist/

[10] Dan Goodin. Crypto flaws in Blockchain Android app sent bitcoins to the wrong address, May 2015. http://arstechnica.com/security/2015/05/crypto-flaws-in-blockchain-android-app-sent-bitcoins-to-the-wrong-address/

[11] Thomas Hühn. Myths about /dev/urandom, November 2016. http://www.2uo.de/myths-about-urandom/

[12] Burt Kaliski. PKCS# 7: Cryptographic message syntax version 1.5. 1998.

[13] Seok Kang. Factors influencing intention of mobile application use. *International Journal of Mobile Communications*, 12(4):360–379, 2014.

[14] David Kaplan, Sagi Kedmi, Roee Hay, and Avi Dayan. Attacking the Linux PRNG on Android. In *8th USENIX Workshop on Offensive Technologies, San Diego*, 2014.

[15] A. Kerckhoffs. La cryptographie militaire (military cryptography). *Sciences Militaires (J. Military Science, in French)*, 1883.

[16] Alex Klyubin. Some SecureRandom thoughts, August 2013. https://android-developers.googleblog.com/2013/08/some-securerandom-thoughts.html

[17] Hugo Krawczyk. The order of encryption and authentication for protecting communications (or: How secure is SSL?). In *Annual International Cryptology Conference*, pp. 310–331. Springer, 2001.

[18] Ilias Leontiadis, Christos Efstratiou, Marco Picone, and Cecilia Mascolo. Don't kill my ads!: balancing privacy in an ad-supported mobile application market. In *Proceedings of the Twelfth Workshop on Mobile Computing Systems & Applications*, p. 2. ACM, 2012.

[19] Greig Paul and James Irvine. Investigating the security of android security applications. In *Proceedings of the 9th CMI Conference on Smart Living, Cyber Security and Trust*, November 2016. http://strathprints.strath.ac.uk/58817/

[20] Greig Paul and James Irvine. Practical attacks on security and privacy through a low-cost Android device. *Journal of Cyber Security and Mobility*, 2016.

[21] Colin Percival. The scrypt password-based key derivation function. 2012.

[22] Israel J Mojica Ruiz, Meiyappan Nagappan, Bram Adams, Thorsten Berger, Steffen Dienst, and Ahmed E Hassan. Impact of ad libraries on ratings of Android mobile apps. *IEEE Software*, 31(6):86–92, 2014.

[23] Suranga Seneviratne, Harini Kolamunna, and Aruna Seneviratne. A measurement study of tracking in paid mobile applications. In *Proceedings of the 8th ACM Conference on Security & Privacy in Wireless and Mobile Networks*, p. 7. ACM, 2015.

[24] Steve Sheng, Levi Broderick, Colleen Alison Koranda, and Jeremy J Hyland. Why Johnny still can't encrypt: evaluating the usability of email encryption software. In *Symposium on Usable Privacy and Security*, pp. 3–4, 2006.

[25] Ryan Stevens, Clint Gibler, Jon Crussell, Jeremy Erickson, and Hao Chen. Investigating user privacy in Android ad libraries. In *Workshop on Mobile Security Technologies (MoST)*, p. 10, 2012.

[26] Michael Sweikata, Gary Watson, Charles Frank, Chris Christensen, and Yi Hu. The usability of end user cryptographic products. In *2009 Information Security Curriculum Development Conference*, InfoSec CD '09, pp. 55–59, New York, NY, USA, 2009. ACM.

[27] Serge Vaudenay. Security flaws induced by CBC padding-applications to SSL, IPSEC, WTLS. . . In *International Conference on the Theory and Applications of Cryptographic Techniques*, pp. 534–545. Springer, 2002.

[28] Yongge Wang and Tony Nicol. On statistical distance based testing of pseudo random sequences and experiments with PHP and Debian OpenSSL. *Computers & Security*, 53:44–64, 2015.

[29] Alma Whitten and J Doug Tygar. Why Johnny can't encrypt: A usability evaluation of PGP 5.0. In *Usenix Security*, Vol. 1999, 1999.

[30] Alma Whitten and JD Tygar. Usability of security: A case study. Technical report, DTIC Document, 1998.

[31] Jeff Yan, Alan Blackwell, Ross Anderson, and Alasdair Grant. Password memorability and security: Empirical results. *IEEE Security & privacy*, 2(5):25–31, 2004.

[32] Li Zhang, Dhruv Gupta, and Prasant Mohapatra. How expensive are free smartphone apps? *ACM SIGMOBILE Mobile Computing and Communications Review*, 16(3):21–32, 2012.

2

"Take It or Leave It": Effective Visualization of Privacy Policies

Prashant S. Dhotre[1], Anurag Bihani[2], Samant Khajuria[1] and Henning Olesen[1]

[1]Center for Communication, Media and Information Technologies (CMI), Aalborg University Copenhagen, DK-2450 Copenhagen SV, Denmark
[2]STES's Sinhgad Institute of Technology and Science, Pune, India

Abstract

As per country law, service providers are obliged to share their business practices with users in the form of a privacy policy. However, considering the complexity of privacy policies, it is questionable to what extent they actually achieve their goal of informing the users. Policies are typically unclear, difficult to understand and time-consuming to read. In this paper, analyzing more than 600 privacy policies of popular websites in India, we present a solution, which can assist users when they are navigating online. The tool performs a semi-automatic analysis and visualization of the privacy policy of the website they visit, and also facilitates access to the reputation of the site. Our solution uses a Naïve Bayes algorithm to classify the privacy policy text across 8 subsections, identified in our previous study. Furthermore, it provides a summarized version of the policy to give users a quick overview of how the service provider handles their personal information. The results show that visual aids can indeed increase the readability of the privacy policy. At the end, we propose a recommended structure of the privacy policy, which can further enhance the user's privacy awareness and understanding of privacy policies.

2.1 Introduction

According to NIST (National Institute of Standards and Technology), Personally Identifiable Information (PII) is defined as:

39

> *"any information about an individual maintained by an agency, including (1) any information that can be used to distinguish or trace an individual's identity, such as name, social security number, date and place of birth, mother's maiden name, or biometric records; and (2) any other information that is linked or linkable to an individual, such as medical, educational, financial, and employment information"* (NIST, 2010).

This definition is quite broad, covering anything that helps to identify and gain more insight about an individual.

Personal information is rapidly gaining social and economic value. According to Tom Cochran *"Personal Information is the currency of the 21st century"* (Cochran, 2013). In India, the big data analytics sector is increasing rapidly, and by 2025 it is predicted to reach a volume of $16 billion (National Association of Software and Services Companies, 2016). The Organisation for Economic Co-operation and Development (OECD) has described an approach to examine the value of personal information that consists of indicators like market capitalization, returns on personal record, market prices, data breach cost, and pay for privacy (OECD, 2013). For example, in the US, the prices for social security number, date of birth, driver's license number, street address, military record are USD 8, USD 2, USD 3, USD 0.50, and USD 35, respectively (OECD, 2013).

There are numerous ways to collect personal information, when a customer interacts with a service provider. The information is often stored and managed in separate units without much coordination, and this has led to a new term called the "Information Silo" (Matthias Reinwarth, 2016). Therefore, the customer identity, its protection and compliance should be at the center of business when providing services.

In the future, providing the right information to the end user and stating the purpose of data collection, will be a major challenge. User awareness and user consent are essential to protect privacy and build trust among service consumers and service providers. Sharing and disclosing of potentially sensitive information should not result in any loss for the customer or the service provider. Both parties should agree on the amount of information disclosure and the conditions for the services that are offered. Hence, organizations should take concrete steps to manage customer information in a way that will value customers as well as the organizations. A Life Management Platform can help users to unite all personal information and provide a set of tools to manage it effectively (OnlyOnce, 2014).

In 2011, India approved a set of rules, the Information Technology Act 2008 (IT Act), for the privacy protection that should be adhered to by customers and service providers (The Gazette of India, 2009). As an important aspect, the IT Act requires any service provider or company that manages personal information to take written user consent before executing any activities. Two sections (43A and 72A) concerning privacy were added. In particular, section 72A speaks about punishment (up to 3 years' custody or a fine of up to Rs. 5,000,000) for a person who causes wrongful gain or loss by revealing information about another person. However, there is some uncertainty in the application of the rules (Ryan, Merchant, & Falvey, 2011).

Similarly, the upcoming European General Data Protection Regulation (GDPR) aims to harmonize the currently fragmented laws for data protection and privacy of EU citizens (The European Parliment and The Council of the European Union, 2016). The GDPR is quite comprehensive and will have a major impact on services and business processes. The idea is to keep personal information protected and focus on explicit user consent for using personal information. The consent form should be phrased in simple words (Dr. Karsten Kinast, 2016). To do this, the user has to be informed about the personal information collection, storage, and sharing. Hence, there is a need for researching how to provide clear and affirmative consent to user information processing. The GDPR also emphasizes the principles of Privacy by Design (Ann Cavoukian, 2011) or Privacy by Default. This implies that privacy should be considered right from the service or product inception. Service providers should only collect the required customer information to satisfy a specific purpose and should discard it when it is no longer required.

Considering the restrictions and responsibilities of the data controller mentioned in Article 23 and 30 of the GDPR, the data controller or processor must define the purpose of the processing of personal information, its category, storage period, and safeguards to prevent unauthorized access or transfer. As a part of the responsibility, each data controller (or controller's representative) shall uphold the record of information processing activities in writing and in electronic form. The record should have details of the data controller, processing purposes, a list of recipients, details of any third countries (in the case of transfer), erasure time limit, etc. Presently, such records are not clearly mentioned in the privacy policy. This is leading to a situation where the user has no clue about what terms and conditions he or she actually gave consent for. Hence, a comprehensive study on the contents of the privacy policy is interesting and necessary. Also, the challenge is to effectively analyze and visualize the privacy policy contents to enhance users' awareness.

Inspired by this, it is important to know the current practices of service providers, especially in India. The strategy of service providers regarding the collection, storage, processing, and sharing of customer information should meet similar conditions as those of the GDPR, as well as the expectations of the customers. In this paper, over 600 privacy policies have been analyzed. Based on this analysis and use of machine learning algorithms, we have developed a Privacy Policy Elucidator Tool (PPET), which can perform automatic classification and summarization of privacy policies. This tool reveals several interesting elements of the privacy policy documents like data collection, the method of collection, security measures, and other important categories. The text classification and machine learning techniques are used to check, which sentences in the privacy policy belong to which category. When a user clicks on this tool, the different sections of the privacy policy are summarized and visualized. It should be noted that the classification and summarization are based purely on the contents of the privacy policy.

Our main research contributions can be summarized as follows:

- Analysis and classification of privacy policy documents into key sections that are of interest to the users.
- Support for increased trust awareness among users by sharing website ratings from WOT (mywot.com, 2006).
- Analysis of how personally identifiable and non-personally identifiable attributes are collected by service providers.
- An automatic analysis and visualization tool to increase the user awareness on key sections of the privacy policies.
- A procedure for automatic summarizing of the text in key sections of the privacy policies.
- A recommended template for the structure of privacy policies of service providers.

2.2 Related Work

The related work presented in this section is divided into two parts, one part dealing with surveys of privacy policies and implementations using machine learning techniques, and another part dealing with privacy-enhancing tools. The focus of our work is on privacy awareness and visualization, not on privacy protection, but both parts contribute to identifying the requirements for our tool.

2.2.1 Survey and Machine Learning-based Methodologies

A college student mockup method was used by (Abdullah, Gregory, & Beyah, 2008) to develop a set of visualization techniques. Based on these visualization techniques, a Firefox extension was developed which aimed at helping the end users monitor their web search activities and increase their awareness about online practices. Most frequently, the visited sites, categorization of disclosed information and time stamp (date and time) were shown using the tool. The work needs to be extended to provide a complete self-monitoring mechanism.

A privacy preserving framework for healthcare data based on K-anonymity and anonymization methods has been proposed by (Chen, Yang, Wang, & Niu, 2012), who elaborated on privacy policy components/elements such as privacy information, allowed set of users, purpose, and methods for use of data. However, more focus should be given to other components like security, consent, third party sharing and retention of data. Manual analysis of privacy policies of popular websites (LinkedIn, Twitter, and Facebook) was conceded to study users' understanding about privacy policies. The conclusion was that Internet users are not able to comprehend what they are agreeing to, when they sign up (Zeadally & Winkler, 2016). Internet users should be educated for their informational privacy. Hence, there is a need to improve the visualization of the content to increase awareness about privacy policies.

A simple visualization tool, "Privacy Pal" (Tucker, Tucker, & Zheng, 2013), has been developed to make users aware of the security and privacy risks associated with third party applications permission granting. The case study shows that Privacy Pal helps to understand the threats to privacy. In the US, a group survey was carried out to check the compliance of 35 frequently used social health websites with the Fair Information Practices (FIP) (Savla & Martino, 2012). An important finding was that the user's privacy is at high risk, if the privacy policy does not comply with the FIP principles, and the privacy policies of healthcare service providers do not support informed decisions.

The completeness of the privacy policy was studied by (Costante, Sun, Petkoviæ, & den Hartog, 2012). The privacy policy was examined against different categories (collection, cookies, security, and sharing) using several methods like Naïve Bayes (NB) classification model, k-Nearest Neighbor (k-NN), Linear Support Vector Machine (LSVM), Support Vector Machine (SVM), and Decision Tree (DT). An add-on helps to browse the text of the privacy policy based on the categories selected by the user, and a grade was assigned depending on the coverage of predefined categories. However, this approach did not consider the semantic values of the defined categories. Also, there is need to show section by section contents of the privacy policy.

2.2.2 Privacy Enhancing Tools

A number of tools have been developed to assist users in not being tracked online. "Privacy Badger" is a browser plug-in developed by Electronic Frontier Foundation (EFF), which can help users to block tracking from advertisers and third parties ("Privacy Badger," 2015). "Lightbeam" is a Firefox add-on that shows the user how the first and third party websites interact and the relationships between the various first and third party websites stored in the user's data using interactive visualizations ("Lightbeam for Firefox," 2012). "Disconnect Me" makes users aware of unsecured connections and hidden requests for the user's personal info and allows the user to block trackers and hackers ("Disconnect Me," 2011). It also provides the ability to control access to the user's personal information.

"Ghostery" enables users to easily detect and control web bugs, which are objects embedded in a web page, invisible to the user ("Ghostery," 2009). Another online privacy protection tool is "MyPermission", which provides a real-time alert to users as soon as any application gets connected, using a single user interface ("MyPermissions," 2014). This app also supports revocation and trust, when the user in online. "Terms of Service; Didn't Read" rates and labels website terms and privacy policies, from "very good" (Class A) to "very bad" (Class E) ("Terms of Service; Didn't Read," 2012). A no guarantee opinion-based project helps the user to understand and analyze terms of service of major websites.

Table 2.1 shows a comparison of these privacy enhancing/awareness tools. They are mostly blended towards third party linking with the services, and they help the users to avoid being trapped by third/unauthorized/untrusted parties.

The survey of related work emphasizes the need for visualization of the content of privacy policies and a better understanding of the privacy policies of service providers. This serves as motivation to develop a tool, which can effectively analyze and automatically visualize privacy policies to empower the end users and increase their privacy awareness.

Table 2.1 Comparison between the difference privacy tools

Tool Name	Functionality	Type of Tool
Privacy Badger	This tool provides assistance to users to block tracking from advertisers and third parties.	Blocking
Lightbeam	Using this tool, the user will be aware of how the first and third party websites interact and their relationships.	Awareness
Disconnect	Unsecured connections and hidden requests for users' personal information are visualized by this tool. Also, this tool allows the user to block trackers and hackers.	Awareness and Blocking
Ghostery	Detecting and blocking of invisible trackers is the main objective of this tool	Blocking
MyPermission	This tool gives users complete control over those apps that access the users' data	Control and Blocking
Terms of Service; Didn't Read (ToSDR)	Using ToSDR, rating and labeling of the terms and privacy policies of major websites can be seen, based on a user community. The ratings cover a range from very good (Class A) to very bad (Class E).	Privacy Awareness
Web of Trust (WOT)	Rating of websites based on user comments/community	Trust Awareness

2.3 Privacy Policy Elucidator Tool (PPET)

The Indian law (The Gazette of India, 2009) mandates websites to have their own policies, which necessitates a thorough study of the privacy policies. In this section, we first discuss the main categories of information in the privacy policies based on our previous work. Using these as a guideline, we then present a detailed description of our proposed tool (PPET), which can analyze and visualize privacy and help the users understand the information collection, storage, processing, and sharing policies regarding the user's personally identifiable and non-personally identifiable data.

2.3.1 Privacy Categories Definition (Core Contents of a Privacy Policy)

To generate a summary, one must gain a general idea about the different types of privacy policies, common contents, and standard industry practices regarding data handling. On these grounds, this research started analyzing privacy policies of different service providers. To obtain the core categories, we considered the common practices discussed in various privacy policies.

These categories were later refined into 8 core categories, considering their importance to the end user (Dhotre, Olesen, & Khajuria, 2016):

1. **Information collection** – It provides the details about the information, the web service providers collect from their users in order to provide their services. This category is further divided into two sections as Personally Identifiable Information (PII) and Non-Personally Identifiable Information (NPII) as defined by NIST (NIST, 2010).
2. **Way of information Collection** – This category elaborates on the way by which service providers collect information from users. For example, registration process, "contact us" forms, etc.
3. **Purpose of the information collection** – This category focuses on the purpose for which information is collected from users. For example, to provide services, to improve the user experience.
4. **Information Sharing** – This category focuses on the other parties like sponsors, affiliates, partners, subsidiaries or law enforcement officials, whom data might be shared with or sold to, and conditions for doing so.
5. **Cookies** – This category provides details about the use of cookies, the type of data collected via cookies, the different types of cookies or other tracking technologies like web beacons used by the service provider.
6. **Policy for Children** – This category explains the data handling policy of the service provider, if the user's age is under 13 or 18, whether they include extra safeguards to protect their privacy, and whether children are permitted to make use of the service.
7. **Information security** – This category provides details about the security mechanisms that the service providers implement to secure user's data. For example, service providers usually implement secure socket layer (SSL), encryption, and physical, managerial or procedural safeguards to protect the user's data.
8. **Others** – This category is for any sentence that does not get classified in any of the above categories of sentences. It is a miscellaneous category that consists of a wide variety of sentences.

2.3.2 General Description of the Tool

The PPET tool is a privacy awareness tool, not a privacy protection tool. It is a semi-automatic tool, which works without any configuration or settings and is developed for users rather than service providers or advertisers. The main goal is to ease the task of reading a privacy policy and make it more engaging to the user. The PPET is developed as a single extension that fetches the analyzed contents of the privacy policy and effectively visualizes them for easy reading. The result is presented to the user in the form of a dashboard, which shows the types of information collection, their methods, security measures, the third party entities and other important privacy-related issues – without the user having to read the complete privacy policy. Using an automatic extraction and summarization method, this tool contributes to empowering users, who will be able to grade a website based on the privacy policy contents.

Figure 2.1 shows how the contents of the privacy policy are visualized in a doughnut chart and divided into different sections. The doughnut chart represents the distribution of sentences of the policy across different sections. Also, a table of statistics is drawn from them. Furthermore, the dashboard presents an extraction-based summary of each section using the word frequency. Finally, the dashboard shows the website reputation obtained from 'Web of Trust' (WOT), a crowd-sourced database (mywot.com, 2006).

The contents of the dashboard are obtained by scraping the text of the privacy policy and parsing its contents. The parsed text gets classified into different sections. Based on our previous survey on the perception of privacy policies (Dhotre & Olesen, 2015), we have identified different sections of the privacy policy, which contain the information that users find interesting. In order to achieve the classification of the text into various categories, text classification and machine learning techniques were employed, as mentioned above. In our case, the classifier was built using a linear algorithm. The reason to select the Naïve Bayes algorithm classification is its computational efficiency (Kibriya, Frank, Pfahringer, & Holmes, 2004), relative ease of implementation over others, and its pervasive availability in many libraries. The advantage of Naïve Bayes over other classification methods is that it gives promising results for textual operations with ease in training the data.

2.3.3 Corpus Design

Privacy policies often have different structures. Some of them are presented in a Q&A format, some are presented as a series of bullet points, and some are divided into different sections, while others can be just monolithic texts. Furthermore, each website has its own unique features and provides different types of services, which requires different types of user data. Thus, there is a fundamental need for a more unified structure of the privacy policy.

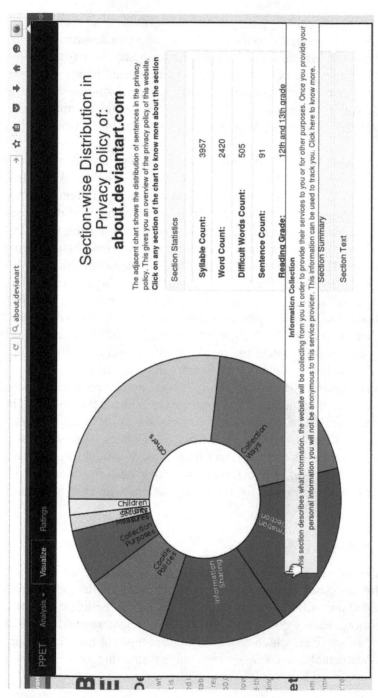

Figure 2.1   Privacy Policy Elucidator Tool (PPET) dashboard (Add-on in action: Privacy policy distribution).

Privacy policies were chosen across the different categories of websites to ensure that each website category was well represented: arts, business, shopping, society, sports, computers, health, home, kids and teens, news, regional, and science. As already mentioned, we identified and examined privacy policies of over 60 websites which are popular in India (Alexa: A Amazon.com company, 1996.). From these, the text was extracted, and each sentence was labelled to create the corpora.

For generating effective classifiers, the corpora need to have a diverse set of sentences. To train a Naïve Bayes classifier, a set of labelled text examples was needed. Hence, for extensive analysis, we have designed our own corpus, consisting of 43,544 sentences considering all categories. Different classifiers were designed for identifying each category. This was done to enable Multi-Category Text Classification, i.e. each sentence can be part of more than one category. So, for each classifier, the corpus contained the same sentences, only their labels varied across different corpora. A positive label was given to the sentences fitting into the respective category, and for other statements, a negative label was assigned. The corpus was further refined and reduced by removing proper nouns, which can form unnecessary informative features. This helped in improving the accuracy of the classifier.

Sentences of each category were stored in the format of Tab-Separated Values (TSV), where the first column contains the sentences and the second column contains the appropriate label for given sentences. Apart from the individual classifiers to identify each category, another classifier was designed to sanitize the inputs given to each of these classifiers. This was required to remove unwanted text from the web pages, such as the header and the footer text. This text typically is not a part of the policy and should not be presented in the dashboard.

2.3.4 Preprocessing

Several activities are performed before the raw text from the corpus reaches the classifier. This is done to optimize the model that is generated after learning. Unnecessary features increase both the temporal and spatial efficiency of the model. The pre-processing techniques involved in this research are: Sentence Tokenization, Removal of Proper Nouns, Stop words removal, Case Removal, Transformation (converting the sentence into an important set of useful words), Stemming, and Lemmatization. This technique will help to chop privacy policies contents into important pieces and form a base for the feature selection.

2.3.5 Privacy Policy Detector

In this study, a classifier is created that could identify a document as privacy policy or non-privacy policy. The input to this was 62 complete privacy policies and 250 other documents. The method involved the development of a web-scraper that starts from the homepage of the website. The web scraper then extracts all the links on the homepage containing the word "terms", "legal", "privacy" and "policy". Next, it opens the webpage and classifies whether the webpage contains the privacy policy. As already explained, the use of Naïve Bayes Classifier as learning algorithm helped to obtain the accuracy for each type of classifier.

2.3.6 Database Description

MySQL is a popular choice of database for use in web applications and is a central component of the widely-used LAMP open source web application software stack (and other "AMP" stacks). LAMP is an acronym for "Linux, Apache, MySQL, and Perl/PHP/Python". Since this implementation makes use of Python and PHP on a Linux machine, MySQL was preferred for development.

The extracted paragraphs are stored in a table. The table contains a column with the website's top-level domain name as the primary key, and there is a column for storing the key components of the privacy policy based on which the paragraphs are classified. The database also contains another table, where the primary key is the website's top-level domain name and there is a column for storing the count of the number of sentences identified in each class during the classification.

2.4 PPET Architecture and Modelling

When an Internet user accesses services offered by a service provider, a privacy policy will – more or less explicitly – be presented to the user. Figure 2.2 represents the architecture to explain the workflow of the privacy policy analysis and visualization process. It consists of mainly three parts:

1. A user who interacts with service providers and PPET,
2. A service provider who provide different services to the end users,
3. The PPET, our system which fetches the policy documents from the service provider and provides visualization to the end user.

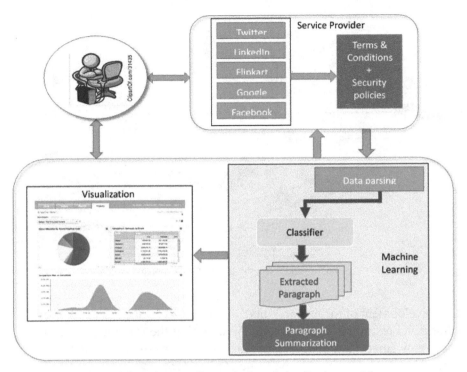

Figure 2.2 Privacy policy analysis and visualization workflow.

The architectural design portrays the developed system. The system consists of a web scraper, which scrapes the privacy policies of different websites. The scraped privacy policies are parsed and passed on to the summarizer system. The summarizer system incorporates the paragraph classifier and paragraph summarizer. Natural language processing can help to detect a pattern. To identify the features and design a model that can perform various tasks, machine learning techniques are helpful. The paragraph classifier is based on the Naïve Bayes machine learning algorithm. It has been trained using a corpus of annotated privacy policies designed by us.

The PPET architecture initially deals with pre-processing of the privacy policy document. Once the text is ready for further processing, this study further uses the Naïve Bayes machine learning algorithm classify and summarize the text for elucidation. Using visual aids, the text will then get visualized for a better understanding of the complicated contents of the privacy policy and to make it more appealing.

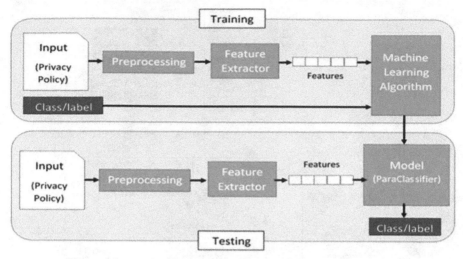

Figure 2.3   The training and testing module of the proposed system.

The development of the learning model is shown in Figure 2.3. It consists of training and testing as the two main parts. During the training phase, the privacy policy (input) is pre-processed (tokenization, stemming, lemmatization, etc.). In this research, we have generated a corpus of 43544 sentences with it. After the removal of stop words, stemming, lemmatization, whatever tokens that are generated are the features.

By default, the Naïve Bayes Classifier uses a simple feature extractor that indicates which words in the training set are contained in a document. The feature vector for the Naïve Bayes classifier is nothing but a "Bag of Words" of such tokens. Each feature set (token and class/label) is converted from values (bag of words) of the pre-processed privacy policy. The feature sets include basic information about the privacy policy, which helps to classify the text into the 8 different sections of the privacy policy (labels). Labels and feature sets are given as input to the machine learning algorithm that gives rise to the ParaClassifier (model).

In the testing phase, the unseen text (new privacy policy) is converted or mapped into feature sets using the same feature extractor as in the training phase. Now, a probabilistic based model (ParaClassifier) will test it and predict the class/label of feature sets.

Initially, the system (S) mathematically consists of three parameters:

$$S = \{I, O, P\} \tag{2.1}$$

where I is the input, O is the output, and P consists of a set of tasks that map input into output. In this proposed method, the input is a text file (privacy policy). Every sentence of the privacy policy will be labelled data. Such labelled data will help to classify the features set into different features.

Mathematically, the features (F) are represented as:

$$F = \{F_1, F_2, F_3, \ldots \ldots, F_N\} \tag{2.2}$$

These features are then classified into 8 different classes (C):

$$C = \{C_1, C_2, C_3, \ldots \ldots, C_k\}; \quad \text{where K} = 8 \tag{2.3}$$

The proposed method involves Naïve Bayes classification method and frequency count based summarization.

2.4.1 Classification

The Naïve Bayes classifier will help to achieve the above goal. This probabilistic model assumes that any feature $F_i \in F$ is independent of other features $F_k \in F$ that gives a mapping value to a class C, see Figure 2.4.

Certain features may be mapped to more than one class. For example, a naïve classifier for two features $\{F_1, F_2\}$ can be defined following some set of structures like

$$\{F_1 \rightarrow C_1\}, \{F_2 \rightarrow C_2\}, \{F_2 \rightarrow C_5\}, \{F_3 \rightarrow C_4\},$$
$$\{F6 \rightarrow \emptyset\}, \ldots \ldots \{F_1 \rightarrow C_8\}$$

In short, the model of classification is C_M and is defined as

$$C_M : F \rightarrow C \tag{2.4}$$

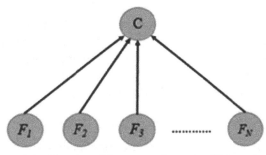

Figure 2.4 Mapping of feature to class using Naïve network.

2.4.2 Summarization and Ranking

Another target of this proposed method is to summarize the text of privacy policy using an automatic text summarizer by measuring the frequency count of a word/token in a sentence. If a token is frequent in a sentence, then probably that sentence is an important topic of the concerned class/label specified in the previous discussion.

During tokenization, tokens are generated for each sentence. Each tuple contains a token and its concerned class/label. Hence, for a privacy policy document, a tuple of tuples will be created. The set of tuples can be represented using 2D matrix T[i][j]:

$$T[i][j] = \begin{bmatrix} S_1W_1, C_1 & S_1W_2, C_2 \cdots & S_1W_m, C_3 \\ S_2W_1, C_3 & S_2W_2, C_8 \cdots & S_2W_n, C_4 \\ \vdots & \vdots & \vdots \end{bmatrix} \quad (2.5)$$

Every entry of above matrix is a pair containing a token and the respective class/label. For an instance, a pair (S_1W_1, C_1) represents the word W_1 from sentence S_1 that belongs to class C_1. The size of T depends on the number of sentences and tokens in each sentence.

The paragraph summarizer subsystem is based on frequency summarizer automatic text summarization algorithms. The Frequency Summarizer tokenizes the input into sentences and then computes the term frequency map of the words.

$$freq[w] = freq[w] + 1; \quad \text{for all words } w \text{ in } S \quad (2.6)$$

Once the frequency dictionary is ready, the maximum frequency is identified. The sentence will be ranked by using the following formula for each word w of sentence i in a document and is expressed in Equation (2.7) as:

$$ranking[i] = ranking\,[i] + freq\,[w] \quad (2.7)$$

The sentences are ranked according to the frequency of the words they contain, and the top sentences are selected for the final summary. In short, words from a tokenized sentence will be input to the summarization. The output is a dictionary of frequency, i.e. the word, its frequency and ranking of sentences. In short, the model of summarization S_M is defined as:

$$S_M : [words\,(w),\ class\,(c)] \rightarrow ranking \quad (2.8)$$

The summarized privacy policies that are obtained after the paragraph summarizer subsystem are integrated and stored in the database.

2.5 Results

This section discusses the important findings. As already mentioned, we identified and examined privacy policies of over 60 websites which are popular in India. The link to privacy policies of these websites was stored in the database along with the site name. Each policy was read to its full extent and the various key parameters for analysis such as the language used, readability, structures and aim of privacy policy, information collection, use of cookies, the method of data collection and its purpose, personal information sharing parties, sharing of personal information, and security measures, were determined.

The Web of Trust (WOT) is a reputation service, which provides reviews and ratings of websites. It works on a unique crowdsourcing approach, which collects reviews and ratings from millions of users across the globe. To retrieve the ratings and trustworthiness of a website, an API provided by WOT were used. Ratings were displayed using C3 Gauge widget. C3 is a D3.js reusable library for charting. Machine learning is involved in generating the dashboard's visualization of the interactive doughnut chart and not in the generation of the reputation.

Figures 2.5 and 2.6 show the behavior of the PPET when the ratings tab is clicked, while visiting the websites of Google and Snapdeal, respectively. This feature helps the end user in making an informed decision whether to trust that particular website. Trustworthiness and child safety represent some of the key parameters to denote the rating of a website. Different colour schemes have been used to represent the level of trust.

Figure 2.5 shows 94% trustworthiness and 91% child safety for Google. This shows that the users trust in Google and believe that its practices and policies are safe, as far as children are concerned. Figure 2.6 shows the ratings for Snapdeal website. In this case, the trustworthiness and child safety is 86%.

This tool has also provided access to our previous work through the Firefox extension (Dhotre et al., 2016). Different pie charts, in total 48 different figures, are generated from the aggregated data obtained after analysis and made accessible through the add-on via a drop down list. The significance of these figures is to tell the users about use and collection of user attributes by the service provider. Each figure represents the distribution of the websites studied based on whether they collect a specific attribute of the users.

Figure 2.7 shows the drop-down list of different figures for 48 different attributes collected by service providers during their interaction with customers. The list contains both personally identifiable (first name, last name,

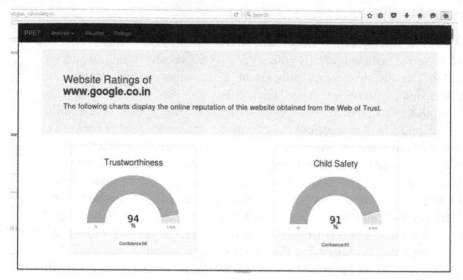

Figure 2.5   Add-on in action: Panel showing the ratings of www.google.co.in

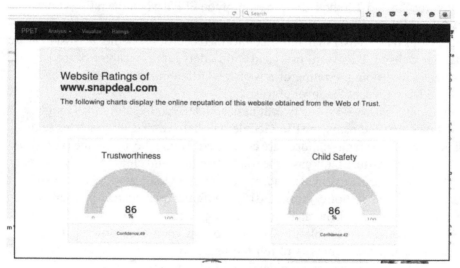

Figure 2.6   Add-on in action: Panel showing the ratings of www.snapdeal.com

address, etc.) and non-personally identifiable attributes (operating system, browser information, etc.). This part of the dashboard is aimed at helping the user know which attributes are important to service providers for making their business effective.

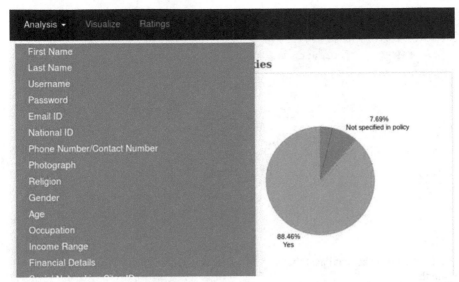

Figure 2.7 Add-on in action: Panel showing the list of attributes (personal and non-personal).

However, this knowledge can also help the user evaluate whether divulging certain information can be thought of as a standard industry practice, thus, to make an informed decision whether that particular service provider's service is really worth the cost of admission.

Any attribute from the given list can be viewed to understand its collection by a number of websites considered in our study. Figure 2.8 represents the use of cookies by different websites. 88.46% of total websites have made a statement on the use of cookies in their privacy policies. However, 8% of the websites said nothing regarding cookies and their use. Less than 4% of the websites clearly mentioned that they do not use cookies. As mentioned earlier, such panels will help the user to understand what the standard industry practice is regarding the collection of email addresses.

The most important outcome of this research is what is displayed in Figure 2.9. This section of the panel in the Firefox Extension presents the user with the doughnut chart of the distribution of sentences of the privacy policy across the core categories. Instead of reading the complete privacy policy document, the proposed method gives a choice to the user to read a particular section of his/her interest. From the doughnut chart, the user can select a section; the right part of the panel will show text only pertaining to that section. Figure 2.9 shows the Information Sharing Section's text from the privacy policy of Flipkart.com, a popular e-commerce website among Indians.

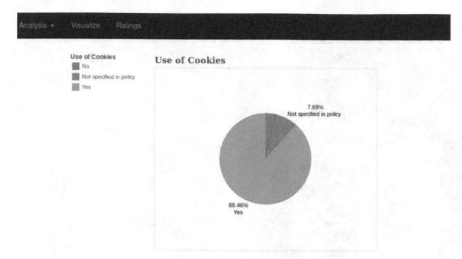

Figure 2.8   Add-on in action: Panel showing the use of cookies among the analyzed privacy policies in this study.

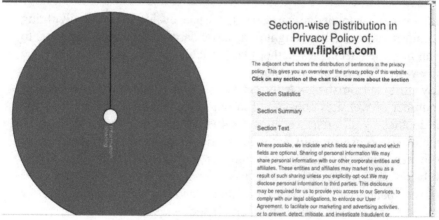

Figure 2.9   Add-on in action: Panel showing the user interested section of the privacy policy (Information sharing from flipkart.com).

The donut chart also provides the user with an overview of the chief contents of the privacy policy that are most relevant to the user. The extensive finding of this research is that the distribution of sentences across the different core categories may be an indicator of the web service provider's concern towards the users' privacy. For example, if the chart indicates that most sentences belong to the section of third party sharing, and there are very few sentences related to information security, it may indicate that the service provider has a low concern regarding the user's privacy. So, we hope this tool will assist the users to make a judgment on their relationship with service providers that benefit the user.

The donut chart visualization not only shows the section-wise text but also gives a summary of those sections. Figure 2.10 depicts the summarized text of the Cookies Section of the Amazon.in the privacy policy. Since most of the privacy policies are lengthy and hard to understand, the PPET will help the user to read any section of the privacy policy that has been made available in summarized fashion. The advantages of the automatic text summarizer is that it will save time from generating the summary manually. The section of the dashboard displays the top 50% sentences in the form of bullet points. It prioritizes the sentences based on the most frequent words from the text.

Furthermore, different statistics are drawn regarding the text from each section, like the sentence count, word count, count of the number of difficult words and the reading ease. The readability is also obtained after obtaining consensus from different readability scoring techniques like Dale-Chall Readability Score, Linsear Write Formula, The Coleman-Liau Index, Automated Readability Index, SMOG Index, The Fog Scale, The Flesch-Kincaid Grade Level and the FleschReading Ease.

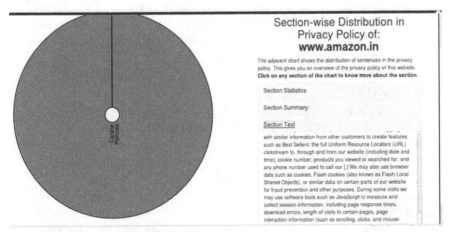

Figure 2.10 Add-on in action: Panel showing the summary of a section of privacy policy (Cookies policy of Amazon.in).

2.6 Recommendations

A meaningful, simple, and understandable privacy policy of service providers helps service consumers to make informed decisions. Providing a standard, transparent privacy policy will be an opportunity to strengthen company brands and stimulate the user's trust. However, privacy policies are very lengthy, difficult to read and offer fewer significant choices to the users (Dhotre et al., 2016).

After analyzing and visualizing the privacy policies of various service providers in this research, we propose and recommend a standard template of privacy policies in India, addressing the following elements:

- **Accessibility and Format**

  A link to the privacy policy should be added to the main pages of the website with appropriate font type and size. Instead of using different writing styles of the privacy policy (bullet point, tabular or plain text), a standard format must be defined for all policies.

- **Readability**

  A privacy policy should use simple, straightforward language to avoid legal or technical jargons. Making effective use of titles and sections (layered format) will guide the users to jump directly to the particular section. Also, the use of graphics or icons will help the users in reading and engaging with the contents of the privacy policy.

- **Objective and Scope**

  In the beginning of the privacy policy, the service provider should describe the objective/purpose of the privacy policy. This should include not only the protection of personal information but also emphasize proper respect to the user privacy. Considering the user practices, the privacy policy should describe the practices of online and offline information.

- **Data Collection**

  There is a strong need to specify the type of information collected (PII, Non-PII) with a complete list of the user attributes. The privacy policy should describe the methods/ways of information collection. This could be resources or technologies (cookies, beacons, etc.) or both. Information retention period should be addressed clearly, e.g. the number of days, weeks, etc.

- **Use of PII**

  Once the information is collected, service providers should elaborate on its use attribute-wise. If the use of PII is unlawful, the user should have the options to withhold consent. The user should be provided with the right access control mechanism to gain/modify the user information with the proper depiction on how to access it.

- **Information Sharing and Disclosures**
  The policy should describe information sharing entities (affiliates or marketing associates) in a prompt and specific way. An information sharing limit must be clearly stated in the privacy policy. Also, the link to those entities must be specified in the policy to gain more details, if the user is interested.

- **Customer Choice and Consent**
  Reasonable efforts should be given when writing statements on the user choices and consent. To obtain the user consent to data collection, sharing or other purposes, clear, explicit, and understandable statements or comprehensive information should be stated in the privacy policy.

- **Security Measures**
  The privacy policy should give a general description of security methods adopted by service providers and their related affiliates. In the privacy policy, there have to be statements on the use of updated security measures time-to-time. Data storage location must be revealed in the privacy policy and must be governed according to the country's law. Effective and efficient encryption methods should be mentioned that protect the user information.

- **Notifications and User feedback**
  For improvement of the privacy policy, there should be an option for the user's input. Service providers should mention the details of the single point of contact if the users want to communicate. Effective notification about privacy policy changes/updates should be used like email.

2.7 Conclusion and Future Work

In this paper, we have proposed a new privacy visualization tool called PPET. The tool has been used to examine, analyze and visualize the privacy policies of more than 600 popular websites in India. Our results demonstrate how the PPET add-on tool can provide help to the users with a section-wise reading of privacy policies.

Different panels of the proposed add-on make it more engaging to the audience to read and understand the privacy policy. Showing the rating of the website in a panel will give confidence to the user to continue doing business and help them build a strong and long-lasting relationship with the service provider. On the other side, the service provider will also get to know the user's perception on their strategies for information collection, sharing, and other important aspects.

The PII and non-PII attributes collection by service providers can be graphically viewed using the PPET tool, providing an indicator to the user for understanding the importance of their personal and non-personal attributes. Using visual aids, a section of the privacy policy can be read effectively without reading the complete text of privacy policy.

Considering the variety of privacy policies in India, we have also presented a recommendation for a template for the privacy policy. The policy should be simple and presentable enough to understand personal information collection, collection methods, data utilization, management and sharing practices, information security measures, and other important activities. A common structure of privacy policy will bring the attention of the users towards privacy policy reading.

We conclude that this add-on will be the first step to spread awareness toward privacy and bring more empowerment to the user. In future, this research and its outcomes will support to design a framework for trust establishment between user and service provider.

References

[1] Abdullah, K., Gregory, C., & Beyah, R. (2008). A visualization framework for self-monitoring of web-based information disclosure. *IEEE International Conference on Communications*, 1700–1707. http://doi.org/10.1109/ICC.2008.328

[2] Alexa:, An Amazon.com company (1996). Alexa – Top Sites in India. Retrieved May 18, 2016, from http://www.alexa.com/topsites/countries/IN

[3] Ann Cavoukian. (2011). Privacy by Design, The 7 Foundational Principles. Information and Privacy Commissioner of Ontario, Canada. Retrieved from www.privacybydesign.ca

[4] Chen, L., Yang, J. J., Wang, Q., & Niu, Y. (2012). A framework for privacy-preserving healthcare data sharing. *2012 IEEE 14th International Conference on E-Health Networking, Applications and Services, Healthcom 2012*, 341–346. http://doi.org/10.1109/HealthCom.2012.6379433

[5] Cochran, T. (2013). Personal Information is the Currency of the 21st Century. Retrieved August 10, 2016, from http://allthingsd.com/20130507/personal-information-is-the-currency-of-the-21st-century/

[6] Costante, E., Sun, Y., Petkoviæ, M., & den Hartog, J. (2012). A machine learning solution to assess privacy policy completeness. *Proceedings of the 2012 ACM Workshop on Privacy in the Electronic Society – WPES '12*, 91. http://doi.org/10.1145/2381966.2381979

[7] Dhotre, P. S., & Olesen, H. (2015). A Survey of Privacy Awareness and Current Online Practices of Indian Users. *Proceedings of WWRF Meeting 34, Santa Clara, CA, USA, Apr. 2015.*

[8] Dhotre, P. S., Olesen, H., & Khajuria, S. (2016). Interpretation and Analysis of Privacy Policies of Websites in India. *Proceedings of WWRF Meeting 36, Beijing, China, June 2016.*

[9] Disconnect Me. (2011). Retrieved September 10, 2016, from https://disconnect.me/about

[10] Dr. Karsten Kinast. (2016). Making Sense of the EU General Data Protection Regulation – YouTube. Germany. Retrieved from https://www.youtube.com/watch?v=MbLFXd_zZuE&feature=youtu.be

[11] Ghostery. (2009). Retrieved September 10, 2016, from https://www.ghostery.com/

[12] Kibriya, A. M., Frank, E., Pfahringer, B., & Holmes, G. (2004). Multinomial Naive Bayes for Text Categorization Revisited (pp. 488–499). Springer Berlin Heidelberg. http://doi.org/10.1007/978-3-540-30549-1_43

[13] Lightbeam for Firefox. (2012). Retrieved September 10, 2016, from https://www.mozilla.org/en-US/lightbeam/

[14] Matthias Reinwarth. (2016). Managing the customer journey – KuppingerCole. Retrieved August 4, 2013, from https://www.kuppingercole. com/blog/reinwarth/managing-the-customer-journey

[15] MyPermissions. (2014). Retrieved September 10, 2016, from https://mypermissions.com/whoweare

[16] mywot.com. (2006). WOT (Web of Trust). Retrieved from https://www.mywot.com/

[17] National Association of Software and Services Companies. (2016). Big data analytics to reach $16 billion industry by 2025: Nasscom – The Economic Times. Retrieved August 10, 2016, from http://economictimes.indiatimes.com/tech/ites/big-data-analytics-to-reach-16-billion-industry-by-2025-nasscom/articleshow/52885509.cms

[18] NIST. (2010). Personally identifiable information. National Institute of Standards and Technology. Retrieved from https://en.wikipedia.org/wiki/Personally_identifiable_information#cite_note-4

[19] OECD. (2013). Exploring the Economics of Personal Data. *OECD Digital Economy Papers*, (220), 40. http://doi.org/10.1787/5k486qtxl dmq-en

[20] OnlyOnce. (2014). THE REVOLUTION IN PERSONAL DATA MANAGEMENT HAS STARTED your Life Management Platform. *OnlyOnce.com.* Retrieved from http://www.onlyonce.com/wp-content/uploads/2014/12/141219-OO-brochure-LMP-V0.61-HARL.pdf

[21] Privacy Badger. (2015). Retrieved September 10, 2016, from https://www.eff.org/privacybadger

[22] Ryan, P. S., Merchant, R., & Falvey, S. (2011). Regulation of the Cloud in India. *Journal of Internet Law*, *15*(4), 13.

[23] Savla, P., & Martino, L. D. (2012). Content analysis of privacy policies for health social networks. *Proceedings – 2012 IEEE International Symposium on Policies for Distributed Systems and Networks, POLICY 2012*, 94–101. http://doi.org/10.1109/POLICY.2012.20

[24] Terms of Service; Didn't Read. (2012). Retrieved September 10, 2016, from https://tosdr.org/about.html

[25] The European Parliment and The Council of the European Union. (2016). REGULATIONS. *Official Journal of the European Union.* Retrieved from http://eur-lex.europa.eu/legal-content/EN/TXT/PDF/?uri=CELEX:32016R0679&from=EN

[26] The Gazette of India. (2009). *Information Technology Act 2008,.* Retrieved from http://meity.gov.in/sites/upload_files/dit/files/downloads/itact2000/it_amendment_act2008.pdf

[27] Tucker, R., Tucker, C., & Zheng, J. (2015). Privacy pal: Improving permission safety awareness of third party applications in online social networks. *Proceedings – 2015 IEEE 17th International Conference on High Performance Computing and Communications, 2015 IEEE 7th International Symposium on Cyberspace Safety and Security and 2015 IEEE 12th International Conference on Embedded Software and Systems, H*, 1268–1273. http://doi.org/10.1109/HPCC-CSS-ICESS.2015.83

[28] Zeadally, S., & Winkler, S. (2016). Privacy Policy Analysis of Popular Web Platforms. *IEEE TECHNOLOGY AND SOCIETY MAGAZINE*, (June), 75–85.

3

A Secure Channel Using Social Messaging for Distributed Low-Entropy Steganography

Eckhard Pfluegel, Charles A. Clarke, Joakim G. Randulff, Dimitris Tsaptsinos and James Orwell

School of Computer Science and Mathematics,
Kingston University, Kingston upon Thames,
London, United Kingdom

Abstract

Free, confidential and uncensored communication between two individuals is an essential requirement for the modern, digital age. Social communication and interaction between users of Online Social Networks (OSNs) enjoy certain security properties, but true end-to-end security cannot be achieved. In this chapter, we devise a secure channel between two OSN Friends using social messaging. Confidentiality is achieved through undetectable communication based on distributed low-entropy steganography, thus implementing a security control for message content vis-à-vis an untrusted OSN provider. A prototype solution, in the form of an Android mobile app, is implemented. This uses two channels of social messaging (Twitter and Google+) to demonstrate the feasibility of the proposed approach. The source code is publicly available, thus making it suitable for use by individuals (and organisations) as a cost-free, secure communication tool.

3.1 Introduction

Free confidential and uncensored communication between two individuals is an essential requirement for the modern, digital age. Advanced and essentially free communication and information sharing functionality is nowadays readily provided by Online Social Networks (OSNs) and used by billions of individuals every day. However, true end-to-end security cannot be hoped for from this form of the communication channel as most OSN terms of use and security practices actually pose confidentiality threats, effectively classifying the OSN provider as untrusted. Using formal security terminology, the OSN provider is referred to as honest-but-curious.

Being concerned about the security of their critical information assets, OSN users will face a range of adversaries, posing as potential threat actors, including the OSN provider. In order to improve this situation, existing OSN Security research has primarily focused on devising controls, ensuring that OSN users can rely on the security goals of anonymity and confidentiality when managing their User Profiles (UPs) and User Generated Content (UGC). The appearance of social messaging applications, which are extremely popular especially in the form of mobile clients, built as front-ends to OSN platforms, introduces a new dimension of communication and the attached security issues are directly related to many aspects of OSN Security.

At this point, it is helpful to introduce two recent proposals for OSN security controls that have been suggested in the literature, which influential for our research. (These are covered in more detail, in the literature review in Section 3.2). They were devised with confidentiality and anonymity in mind, and also taking into account that all secret communication that is routed through a centralised OSN needs to be *undetectable*, even if routinely filtered through statistical or semantic analysis – c.f. the terminology of a socially constrained channel in Section 3.1).

Firstly, confidential communication techniques for untrusted OSNs have been proposed [3], and in particular using low-entropy steganography. This implements a security control against threats to UGC vis-à-vis an untrusted OSN provider.

Secondly, Virtual Private Social Networks (VPSNs) have been introduced [8] and further developed [9] as a powerful conceptual framework for leveraging existing OSN infrastructure to provide UP anonymity, while still providing the full functionality of the OSN. The VPSN architecture fits traditional, web-based OSN platforms.

The research in this chapter extends [3] by using multiple OSNs, and moving focus away from traditional OSN platforms to mobile architectures. The VPSN approach could be seen as complementary to our solution, especially our previous work [5, 6]. This aspect is further discussed in the conclusion section of this chapter.

3.1.1 Outline of Proposed Method

The aim is for two OSN users, Alice and Bob, to exchange a secure message by using social messaging. This main requirement is that the content of the message is not able to be deciphered by any third party, including the OSN provider (e.g. Facebook, Twitter, Google+), which could hypothetically take the role of the 'man-in-the-middle' in an attack of the same name. An additional requirement is that the message is undetected, in the sense that it is difficult for any third party to identify all and only those elements of communication necessary to decrypt the message. To summarise, Alice and Bob need end-to-end security, which is not normally implemented in these scenarios.

The standard approach to implement message security is based on encryption. Regardless of whether symmetric or asymmetric encryption is used, key exchange is a critical task that needs to take place 'out-of-band' before the actual encryption process. When attempting to apply this to our scenario, Alice could send a secret key to Bob using social messaging, and she could then transmit an encrypted message using a different social messaging channel. This would be secure under the assumption that two different OSN providers do not co-operate. The problem with this approach is that the presence of a ciphertext in the social messaging channel will stand out from "normal" plain text conversations, using natural language. This could potentially raise suspicion and would draw attention to Alice and Bob in the presence of traffic analysis.

Thus, under these circumstances the, social messaging can be regarded as a *socially constrained channel* in which messages containing deviations from natural language patterns will be noticed and hence need to be avoided. The OSN provider may eavesdrop on all communication and in particular any "unusual" content such as encryption (if tolerated by the social network) or artificial messages used for steganography will also raise suspicion.

The previously proposed solution [3] uses the socially constrained channel as a side-channel in order to send an inconspicuous cover message text t, which to any third party would be regarded as a reasonable and genuine communication. A second channel can be characterised as socially *un*constrained. This can be used to transmit the actual ciphertext, obtained by encrypting the plaintext message with a symmetric key, which is obtained by using t as a seed, e.g. by applying a secure hash function. This protocol is in effect using information hiding (referred to as low-entropy steganography). For this to work, Alice and Bob need access to each other's OSN account names, and the cryptographic parameters used such as the specific cipher and hash algorithm.

It is worth noting that, both VPSN [8] and undetectable communication architectures [3] are instances of the two-channel communication genus – the first communication channel being a socially constrained OSN channel, the second socially unconstrained channel a TCP/IP connection to an XMPP server [8], or an external file server such as Dropbox [3].

Below, these approaches are generalised, to propose a protocol making use of one 'regular' (socially unconstrained) channel and *n* socially constrained channels. This proposal has a number of benefits that are explored in Section 3.3.

3.1.2 Research Contributions

The contribution of this chapter is twofold, comprising a theoretical result and its practical evaluation through the implementation of a prototype solution.

The first contribution is the proposal for a secure channel, achieving confidentiality through undetectable communication based on distributed low-entropy steganography using social messaging.

The second contribution is the development and release of a prototype implementation, in the form of an Android mobile app, using on communication using multi-channel mobile instant messaging through Twitter and Gmail.

3.1.3 Chapter Organisation

This chapter is organised as follows: in Section 3.2, the relevant previous work is reviewed, and build on these ideas to shape the research question central to this chapter. In Section 3.3, the proposed system architecture is presented, followed by a description and evaluation of our implementation in Section 3.4. Some concluding remarks are provided in Section 3.6.

3.2 Previous Work

The proposed approach builds on several key papers [1, 3, 8, 9] in the field. In this section, after a general discussion of OSN security issues, the main results of these papers are reviewed, and an indication is given of the direction of the follow-up research [5], which precurses the work described in this chapter.

There are a wide range of security and privacy threats to OSN-related assets, as illustrated in this section. From the point of view of OSN providers, they need to protect user data assets against attacks from both external attackers and insiders (other OSN users). Examples of OSN-internal (user) attacks are:

- False account registrations – providing wrong personal information (such as name, town, gender).
- Identity masquerading – malicious theft and consequent misuse of another OSN user's identity to commit a crime.
- Use of their platforms for illegal purposes – the use of advanced OSN communication functionalities for discussing, planning or executing criminal activities.
- Malware – malicious software is infiltrated into the OSN system, for potential downloading by other users.

Typical OSN-external attacks are:

- Web application attacks – the OSN's web interface is targeted by typical attacks such as Denial of Service (DoS), Buffer Overflow, Cross Site Scripting (XSS) or SQL Injection.
- Data Breach – user-related information is disclosed to an external party, compromising the system.
- Account Compromising ("hacking") – user account information or any other user data is disclosed to unauthorised parties.

We can assume that the OSN provider provides integrity, both in terms of data and origin integrity. However, as already mentioned, other OSN users might betray their peers through various attacks. Even more crucially, the OSN Provider has access to all UGC and user account data and is able to intercept and potentially modify all messages sent between users.

The main OSN-provider threats to UGC are as follows:

- Data Exploitation – an OSN host may impose the right to use UGC for commercial or marketing purposes, without the need to consult, or compensate the user.
- Data Censorship – an OSN host may impose the right to modify or remove UGC for reasons of censorship or violation of terms and conditions.
- Data Sanitisation – OSN hosts may sanitise user data prior to publication, in order to protect themselves and other users from malware.

Finally, digital surveillance is reaching a worrying scale, and the Internet Service Provider (ISP) needs to be added to the list of adversaries. The ISP adversary is as capable as the OSN Provider adversary, but additionally, he can monitor the Internet traffic between any OSN users.

A number of cryptographic techniques have been suggested in the literature to address the security goals of confidentiality, integrity and availability in the presence of various adversary types. Encryption was suggested in early work [1, 10, 12, 14–16, 19]. Steganography, proposed in [2, 3, 13], was recognised as being a more suitable technique due to its hiding properties.

Security controls should be transparent to the OSN host and users, and should not impair the performance of the OSN. These additional requirements lead to the idea of implementing security based on devising a novel OSN system architectures.

Decentralisation prevents UGC to be stored and controlled on a centralised server [4, 21] but this has proven to be difficult to implement.

The work described in the remainder of this section is based on two-channel communication architectures (or more generally, n channels as employed by this chapter), which is a promising approach, combining desirable security and usability features with ease of implementation.

However, in all of these approaches, the potential threat of traffic analysis is a problem. Traffic analysis refers to a process of message or data intercepting, from which pattern based information (e.g. trends in communication times, data size etc.) can be deduced, even if the messages or data are encrypted.

3.2.1 UP Anonymity

In [8], the authors conceptualise the notion of a VPSN, in which the entities of a traditional Virtual Private Network (VPN), are mapped to elements of an OSN. The OSN is deemed analogous to a public untrusted infrastructure and the users are analogous to network devices. The objective is to implement a means of secure communication between users, via the OSN, that delivers the security goals of a VPN, hence the term VPSN.

A VPSN enjoys a great number of properties, and an ideal implementation would result in a system offering the full feature set of the underlying OSN, as well as additional security. The essential VPSN characteristics as introduced in [8] can be summarised as follows:

- A VPSN is virtual, since it utilises the functionality of an existing OSN and is by nature an OSN itself. In particular, a VPSN inherits security mechanisms from the OSN. These include authenticated access and authenticated HTTP connections between users. However, being a private entity, user profile information can be hidden from any non-intended audience (crucially, this includes the OSN provider).

- A VPSN is hidden to users that are not part of it, as well as the OSN provider. In addition, it is transparent to its users and hence provides the desirable combination of security and usability.
- A VPSN solution should allow members to connect to multiple VPSNs [or OSNs???]. The impact of a VPSN on the OSN's performance should be negligible and VPSN users should still have access to the full functionality of the OSN.

VPSN research addresses a central problem of how to make the public facing elements of a Facebook user profile, private to all but intended friendship groups. The solution relies on users creating Facebook accounts, with fake user profile credentials. By default, the publicly accessible profile information displays the fake credentials. True profile values are revealed to legitimate VPSN members via privately shared XML lookup tables that render dynamically when the user profile is visited. This idea is implemented as a Firefox plugin called FaceVPSN as further detailed in [9].

Virtual Private Social Networks were primarily introduced with user profile privacy and anonymity in mind. This was achieved by restricting the availability of profile data. A benefit of traditional VPNs when operating in so-called tunnelling mode is that they hide the endpoints of the communication link from eavesdroppers. It can be noted that user profile anonymity in a VPSN could be seen as an analogy of the existence of hidden endpoints of a VPN tunnel between nodes in the network.

To our knowledge, the only VPSN implementation described in the literature is given in [9], where the authors describe FaceVPSN, implemented as a Firefox web browser plug-in. An alternative VPSN architecture was given in [5], although the contribution remained conceptual.

3.2.2 UGC Confidentiality

The idea of [1] and later [3] is to disguise UGC through a specific form of steganography, which makes it "socially indistinguishable" and hence difficult to detect (under certain assumptions). The chapter gives a formal definition of communication models, enabling undetectable communication in OSN, based on two schemes: a high-entropy model using traditional steganography techniques for cover modification, selection or synthesis and the low-entropy model, based on specialised techniques for the usage in OSNs.

The first scheme adopts a conventional theme of hiding a secret payload within a cover object, which is then simply transmitted as carrier object by the sender, via the OSN to a recipient. This could involve a secret key, in order to enhance the security of the scheme, which needs to be exchanged out-of-band before the message can be understood by the recipient.

The second scheme is described as an information sharing scheme that is provably undetectable within the OSN. In this scheme, it is assumed that a prospective cover-object has no capacity for embedding a secret as a payload (e.g. if an OSN such as Twitter applies a constraint that limits the capacity of the cover-object). Therefore, rather than containing a payload, the cover-object (which may be a keyword), represents the location of where a secret can be accessed. Prior to exchanging information via this scheme, a sender and receiver pre-agreed keywords that are mapped to online locations that are hidden by way of a URL shortening service. In this approach, the pre-shared keywords that are mapped to a shortened URL, represent a stego-key. Conceptually, when a recipient receives a particular keyword in a message posted to the OSN by the sender, they match the keyword to the pre-agreed location, which they visit to access the secret.

The general concept of how the scheme works is as follows:

1. The sender visits a storage location to deposit a secret. For each recipient a separate folder is created, in which a separate copy of the secret is stored.
2. The sender uploads the storage URL and keyword mapping index to the URL mapping service, and subsequently posts an appropriate keyword to the OSN.
3. A participating recipient accesses the innocuous keyword, derives the mapping index and visits the URL mapping service to derive the URL of the storage location for the secret.
4. The recipient visits the storage location to access the secret.

The example presented is based on the use of Facebook as the OSN, TinyURL as the URL shortening service and Dropbox as the storage server. It is assumed that links between all nodes are secured via HTTPS and that encryption keys and stego-keys are exchanged via secure OOB channels. The secret is ultimately protected by the use of encryption.

3.2.3 Distributed High-Entropy Steganography Approach

It is seen that both reviewed approaches use a two-channel communication architecture – the first communication channel being the OSN channel, the second channel a TCP/IP connection to a XMPP server in the case of FaceVPSN, and an external file server such as Dropbox in [3].

Revisiting the notion of socially constrained and unconstrained communication channels, we identify the OSN channel as the socially constrained, and the second channel as the socially unconstrained channel. The OSN provider may eavesdrop on all communication, and in particular any "unusual" content such as encryption (if tolerated by the social network) or artificial messages used for steganography will raise suspicion.

An alternative architecture based on a distributed communication protocol using n channels has been proposed in [5]. This approach combines steganography with another fundamental cryptographic technique: secret sharing. The idea of secret sharing is to divide given data (the secret s) into n parts (the shares) in such a way that knowing at least m shares allows for reconstructing s. In an ideal secret sharing scheme, knowledge of less than m shares will not reveal any information on s. A secret sharing scheme with parameters m and n satisfying the aforementioned properties is also called a (m, n) threshold scheme. A popular scheme is based on polynomial interpolation, introduced by Shamir [18].

The approach of [5] effectively yields a high-entropy steganographic technique with additional security and robustness due to the nature of the secret sharing scheme. The approach can be described as follows. First, an (m, n) secret sharing scheme is applied to the plaintext message, splitting it into n shares. High-entropy steganography is used to hide the individual shares in a suitably crafted carrier-medium, thus creating a stego-medium. Each stego-medium (which to the OSN takes the form of generic and ordinary UGC) is then sent on the individual OSN channel.

The recipient can access the original message by retrieving a subset of m messages, extracting the shares from the stego-medium, and combining them in order to reconstruct the message.

Apart from the need to carry out steganalysis in order to detect individual secret payloads, the correct combination of shares would have to be identified using brute-force searching across different users and different messages per user. In addition, the particular secret sharing scheme would have to be known in order to reconstruct the secret message m from the shares. These tasks seem intractable when considering the potential number of OSNs, users, messages, steganographic and secret sharing scheme combinations, which informally justifies the claim of additional security.

3.3 Proposed Architecture

The key contribution of this chapter is given in this section. It consists of a cryptographic protocol and an OSN system architecture.

The proposed protocol uses a suitably adapted secret sharing scheme with random shares, inspired by the scheme presented in [22]. The provided high-level description hides the complex details that would be required for the specification of a robust real-world security protocol. As already explained in Section 3.2, this is a low-entropy steganography approach as introduced in [3], with improved payload capacity compared to [5].

The system uses a distributed architecture (see below for a more detailed discussion) inspired by [5], in turn improving on the two-channel approach of [3] by achieving additional security and redundancy.

In this solution, undetectable communication is achieved by using n socially constrained and one additional, socially unconstrained channel.

In order for the proposed scheme to be operational and secure, some additional assumptions are required, as follows. Alice and Bob have accounts with n OSNs, on each of which they are mutual friends. They also have (joint) access to an external file server. The OSN authentication mechanism is trusted. This is a reasonable assumption, as a compromise would threaten trust in the OSN. The individual OSNs do not co-operate, i.e. they respect the confidentiality of user data with regards to external entities. We again deem this expectation to be reasonable.

In order to send a secret message m, Alice and Bob proceed as follows – it is assumed that both Alice and Bob have previously exchanged a secret key K_{AB} and suitable parameters g and p where p is a large prime number and g a primitive root modulo p. A secure hash function H is also required. The integers k and n are as in the previous section. Figure 3.1 provides a diagrammatic overview.

1. Alice chooses n short texts $t_1, \ldots, t_n$. She then sends t_i to Bob, using OSN_i.
2. She also chooses a random polynomial $f(x)$ of degree k, satisfying $f(0) \neq 0$. The session key is then $K = (H(K_{AB}, f(0))$.
3. Furthermore, she computes $h_i = H(t_i)$ and $g_i = g^{hi} \bmod p$. She finally sends $c = E_K(m)$, $[g_1, \ldots, g_n]$ and $[f(h_1), \ldots, f(h_n)]$ using the socially unconstrained channel.

Message reconstruction then works as follows:

1. Bob selects k out of the n OSNs and reads his messages, hence obtains a set of k short texts t_{ij} $(j = 1, \ldots, k)$. He computes the corresponding h_{ij} and g_{ij}.

2. Listening on the socially unconstrained communications channel, he receives c and the additional data. He uses this to identify the corresponding $f(h_{ij})$.

3. Using polynomial interpolation, he constructs $K = H(K_{AB}, f(0))$. Finally, he obtains the original message by decrypting c.

We observe that for $k = n = 1$, we will obtain the undetectable communication scheme of [3]. If $n >= k > 1$, two additional benefits arise from our scheme:

- The value n–k measures the robustness against the unavailability of OSNs. In particular, this implies a resistance against an account closure attack of a leveraged OSN, potentially in response to suspicious user activity.

- The value k is a threshold value for the number of OSNs that would have to cooperate in order to combine individual shares of the short text t. Any number of OSNs less than this value can provably not gain any information about t, due to the properties of the secret sharing scheme.

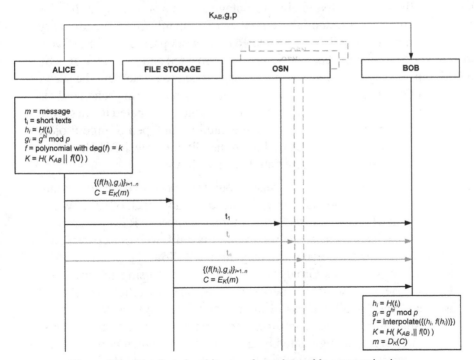

Figure 3.1 Distributed architecture for undetectable communication.

3.4 Implementation

In this section, the demonstration proof-of-concept is described.

The system is realised as an Android mobile app, implemented in the Java programming language. Running within the Android operating system, the application uses the application programming interfaces (API) of the OSNs in order to access the needed communication functionality. Implementing the solution in the Android operating system (OS) is a choice that improves security, in part due to the fact that Android is an Open Source development platform. Because of Google's focus on privacy, the Android OS includes several security controls related to protecting applications and user data, and the fact that the OS itself is built on the well-known Linux kernel is one of its strongest security features. The openness of Android also makes it easier to develop applications, and by utilising the built-in permissions feature, the users of the application can themselves examine what type of data the system will be able to access.

Currently, the implementation uses Twitter and Gmail, but it would be possible to extend the number of available channels by adding additional APIs. Using the APIs of these OSNs is a big advantage in terms of security, as it helps a developer implement potentially complicated authentication functionality with fairly simple code. All the logic related to network protocols, HTTPS and authentication has already been designed, implemented and soundly tested by the OSNs themselves. An open source library is used to implement the secret sharing functionality, which is based on Shamir's Secret Sharing scheme. This library has been reviewed by cryptographers, who have evaluated it as a secure solution. The mobile application is implemented as an Open Source project and the source code is publicly available on the GitHub website [17].

The basic data flow of the application is as follows:

1. Prior to exchanging secret messages, both users need to have an account with Gmail and Twitter, and they also need to know each other's Twitter username and Gmail address. Furthermore, they need to follow each other on Twitter, which is due to the way Twitter is built, where users can only privately message someone who they already follow.

2. The user logs into his Gmail and Twitter account using his own credentials. He then writes a message string, and specifies the Twitter username and Gmail address of the recipient. The system then applies secret sharing to the message string, thus creating two different shares. The shares are then augmented by a unique suffix, which consists of an application-specific identifier and the globally unique ID (GUID) of the Android device. This suffix is added in order to make it possible for the

application to identify shares when it is looking for incoming messages. These augmented shares are each distributed using the OSN channels, with one share being sent as an email, and the other being sent as a private Twitter message.

3. The recipient, who is logged in with his own Gmail and Twitter credentials, then tells his application to look for a message. The application accesses the Gmail and Twitter inbox of the user, and looks for the augmented shares, by looking for the application-wide keyword that was added to the share suffix before sending. After identifying the potentially multiple shares the application uses the Android GUID to identify shares coming from the same sender. By then comparing the date and time of receiving a message the application can identify shares that belong together. When they are found, the application can join the shares to recreate the original message string from the other user.

The following figures depict the Graphical User Interface of the mobile app, offering the user a choice between sending and receiving a message. Sending a secret message only requires entering the text and recipient addresses,

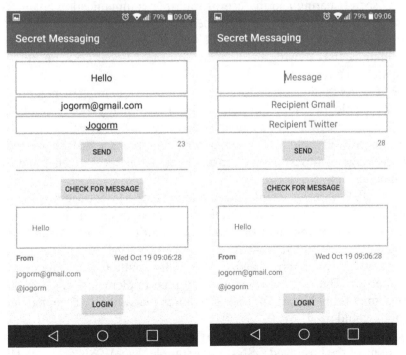

Figure 3.2 Sending and receiving a message.

if the user has previously configured his Gmail and Twitter account details. The secret message is received by the two communication channels and automatically reassembled to obtain the plaintext message.

The application has been evaluated in terms of its computational overhead and no significant delays occur when sending messages of a reasonably small size compared to using a standard messaging app. Currently, the message size in this tool is limited to a small number of characters, due to limitations imposed by the Twitter API – this is unfortunately less than 140, the number of characters in a Twitter tweet. The downtime of global OSNs such as Gmail and Twitter is typically limited to a couple of hours per year, and there are thus no real danger of availability issues for this application. This solution is currently only using two communication channels, and the security could be improved substantially by using other OSNs as well. Using a third channel simply for redundancy reasons could be a valuable improvement.

3.5 Conclusion

This chapter describes the design of a secure channel, based on social messaging, secret sharing and the use of multiple communication channels for message transmission. It also describes the implementation of a social messaging Android app using the Twitter and Gmail APIs as a proof-of-concept solution, thus moving focus away from traditional OSN platforms to mobile architectures.

It would be relatively straightforward to enhance this implementation to yield a fully usable, robust system with richer functionality. For example, a further enhancement could be provided by the automatic suggestion of cover messages that fit well with the online personae and message history. Indeed, plausible message histories could themselves be automatically generated to better camouflage the cover message.

To date, OSN security architectures and systems do not seem to have attracted a lot of attention for commercial exploitation, despite their readiness for robust implementations and the resulting security benefits. A range of issues might be underlying reasons:

Implementing a truly robust and user-friendly solution would require a substantial effort. The existing research prototype implementations demonstrate proof of concept but are lacking of features that a widely accepted and adopted tool would need.

Some of the systems, depending on their architecture, may violate the terms and conditions of the leveraged OSNs. For example, FaceVPSN achieves UP anonymity by using fake user profiles. In [11], Facebook report that the total percentage of "fake" or duplicate user accounts that violate their terms and conditions, is estimated to be 7% (116 million users out of a total 1.65 billion monthly active users). Whilst it might be impractical for an OSN to detect such fake accounts, it might be difficult for a commercial VPSN solution to become viable and remain legally unchallenged.

There has been a change in use patterns, with users increasingly relying on mobile devices and the use of mobile apps for social networking, in particular, mobile instant messaging apps, are one of the most heavily used mobile apps [20]. Clearly, there is a need for more research in this area, in particular in the design of mobile-based security solutions.

We argue that OSN security architectures are a powerful concept – yet to be discovered for future use by individuals and organisations alike, and that this work is a step in this direction.

References

[1] Beato, F., Kohlweiss, M., & Wouters, K. (2011). Scramble! your social network data. In Privacy Enhancing Technologies (PETS) (pp. 211–225). https://doi.org/10.1007/978-3-642-22263-4_12

[2] Filipe Beato, Iulia Ion, Srdjan Čapkun, Bart Preneel, and Marc Langheinrich. 2013. For some eyes only: protecting online information sharing. In Proceedings of the third ACM conference on Data and application security and privacy (CODASPY '13). ACM, New York, NY, USA, 1–12. DOI=http://dx.doi.org/10.1145/2435349.2435351

[3] Beato, Filipe, Emiliano De Cristofaro, and Kasper B. Rasmussen. "Undetectable communication: The online social networks case." Privacy, Security and Trust (PST), 2014 Twelfth Annual International Conference on. IEEE, 2014.

[4] Buchegger, S., & Schi, D. (2009). PeerSoN: P2P Social Networking – Early Experiences and Insights. In Proceedings of the Second ACM EuroSys Workshop on Social Network Systems (pp. 46–52). ACM. https://doi.org/10.1145/1578002.1578010

[5] C. A. Clarke, E. Pfluegel and D. Tsaptsinos, "Confidential Communication Techniques for Virtual Private Social Networks", DCABES 2013: The 12th International Symposium on Distributed Computing and Applications to Business, Engineering and Science, Kingston University, Sep. 2–4 2013.

[6] C. A. Clarke, E. Pfluegel and D. Tsaptsinos, "Enhanced Virtual Private Social Networks: Implementing User Content Confidentiality" in ICITST 2013: The 8th International Conference for Internet Technology and Secured Transactions, London, UK, Dec. 9–12 2013.

[7] Clarke, Charles A., Eckhard Pfluegel, and Dimitris Tsaptsinos. "Multi-channel overlay protocols: Implementing ad-hoc message authentication in social media platforms." Cyber Situational Awareness, Data Analytics and Assessment (CyberSA), 2015 International Conference on. IEEE, 2015.

[8] Conti, Mauro, Arbnor Hasani, and Bruno Crispo. "Virtual private social networks." Proceedings of the first ACM conference on Data and application security and privacy. ACM, 2011.

[9] Conti, Mauro, Arbnor Hasani, and Bruno Crispo. "Virtual private social networks and a facebook implementation." ACM Transactions on the Web (TWEB) 7.3 (2013): 14.

[10] De Cristofaro, E., Soriente, C., Tsudik, G., & Williams, A. (2012). Hummingbird: Privacy at the time of Twitter. In Proceedings – IEEE Symposium on Security and Privacy. https://doi.org/10.1109/SP.2012.26

[11] Facebook Inc., "Facebook – Financials – SEC Filings," fb.com, 2016. [Online]. Available: https://investor.fb.com/financials/sec-filings-details/default.aspx?FilingId=11342580. [Accessed: 20-October-2016].

[12] Feldman, A., & Blankstein, A. (2012). Social networking with frientegrity: privacy and integrity with an untrusted provider. In Proceedings of the 21st USENIX conference on Security symposium (Vol. 12, pp. 31–31). Retrieved from https://www.usenix.org/system/files/conference/usenixsecurity12/sec12-final67.pdf

[13] Ion, I., Beato, F., Capkun, S., Preneel, B., & Langheinrich, M. (2013). For some eyes only: Protecting online information sharing. In CODASPY 2013 – Proceedings of the 3rd ACM Conference on Data and Application Security and Privacy (pp. 1–12). https://doi.org/10.1145/2435349.2435351

[14] Lucas, M., & Borisov, N. (2008). FlyByNight: Mitigating the Privacy Risks of Social Networking. In Proceedings of the Seventh ACM Workshop on Privacy in the Electronic Society (pp. 1–8). https://doi.org/10.1145/1456403.1456405

[15] Luo, W., Xie, Q., & Hengartner, U. (2009). FaceCloak: An Architecture for User Privacy on Social Networking Sites. In International Conference on Computational Science and Engineering (Vol. 3, pp. 26–33). https://doi.org/10.1109/CSE.2009.387

[16] Malik, S., & Sardana, A. (2011). Secure Vault: A privacy preserving reliable architecture for Secure Social Networking. In Proceedings of the 2011 7th International Conference on Information Assurance and Security, IAS 2011 (pp. 116–121). https://doi.org/10.1109/ISIAS. 2011.6122805

[17] "Secretmessaging Open Source Project", github, https://github.com/ jogorm/Secretmessaging [Accessed: 20-October-2016].

[18] Shamir, A. (1979). How to share a secret. Commun. ACM, 22(11), 612–613. https://doi.org/10.1145/359168.359176

[19] Sorniotti, A., & Molva, R. (2010). Secret interest groups (SIGs) in social networks with an implementation on Facebook. In Sac 2010 (pp. 621–628). https://doi.org/http://doi.acm.org/10.1145/1774088. 1774219

[20] Statista Inc., "Mobile messaging users worldwide 2014–2019," statista. com, 2016. [Online]. Available: https://www.statista.com/statistics/ 483255/number-of-mobile-messaging-users-worldwide/. [Accessed: 20-October-2016].

[21] Yeung, C. A., Liccardi, I., Lu, K., Seneviratne, O., & Berners-Lee, T. (2009). Decentralization: The Future of Online Social Networking. In W3C Workshop on the Future of Social Networking Position Papers (Vol. 2, pp. 1–5).

[22] Zhao, J., Zhang, J., & Zhao, R. (2007). A practical verifiable multi-secret sharing scheme. Computer Standards and Interfaces, 29(1), 138–141. https://doi.org/10.1016/j.csi.2006.02.004

4

Computational Trust

Birger Andersen, Bipjeet Kaur and Henrik Tange

Technical University of Denmark, Ballerup Campus (DTU)

4.1 Introduction

In the world of Internet of Everything, intelligent autonomous systems and cloud-based knowledge management systems, data is gathered from several sources such as application, sensor devices, documents from several sources which contain information used to develop services, etc. These services may be regarded as having a degree of trustworthiness depending on the trustfulness of the sources fused together for obtaining the output of the service. To understand this whole concept of what a trustworthy service is, we need to understand several keywords and concepts which are described next.

Internet of Everything (IoE) means everything and anything connected to the internet for intelligent autonomous systems or decision making. IoE is including internet of PCs, gateways, things, servers etc. – anything which is able to send messages to the anything on the internet. It brings together people, process, data and things to make networked connections more relevant and valuable than ever before turning information into actions that create new capabilities, richer experience and unprecedented economic opportunity for businesses, individuals and countries. Figure 4.1 explains the concept of IoE that includes data which results in leveraging data into useful information for decision making, things which are physical devices and objects connected to the Internet and each other for intelligent decision making along with processes delivering right information to the right person who want to collect relevant and valuables ways.

In general, "Conceptual trust" term can be defined as both a personality trait and a belief and both as a social structure and behavior intention. Conceptual trust also takes into account personal view on the fact that depends on like

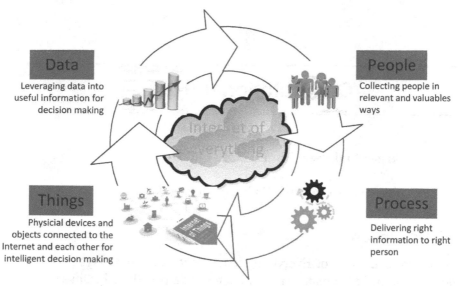

Figure 4.1   Concept of Internet of Everything (IoE).

or dislike of person, environment or thinking, whereas computational trust is a result of logical algorithm running to accumulate the trust value of individual sources of information to develop the output of service. It also takes into consideration the security of system while assigning trust values to the parameters used in the algorithm.

4.2 Trust

Increasing use of Internet in diverse fields has a demanding association in security and trust of services. The trust on the services cannot exist independent. It is greatly depended on the security of system at each layer. Several definitions of trust has been coined to clear the sky of trustworthiness.

In a social context, trust has several connotations [1]. Definitions of trust typically refer to a situation characterized by the following aspects: One party (trustor) is willing to rely on the actions of another party (trustee); the situation is directed to the future [2]. In addition, the trustor (voluntarily or forcedly) abandons control over the actions performed by the trustee. As a consequence, the trustor is uncertain about the outcome of the other's actions; they can only develop and evaluate expectations. The uncertainty involves the risk of failure or harm to the trustor if the trustee will not behave as desired [3].

Quite similar to the term Knowledge – it always depends on the specific environment and field of research and application, what you understand within the term within "Trust".

Usually, we have a high trust in man-made technology – from cars to airplanes, from computers and buildings to space shuttles. As long as, they are working properly, we most of the time do not consider whether we trust them or not. Only in case they stop working in their usual behaviors, the question of trust comes up. The trust in IT systems is becoming even more important, as today's people rely on IT more than ever before. Besides the usage of IT in every part of our lives, security has been given quite an importance with the Internet domain. Now the question of the trust into content from the Internet arises and has to be handled crucial. The main threat of trust does not arise from the viewing the contents of a webpage or retrieving information but is from the download and linked mechanism behind this. When you download a file on your computer, it is highly vulnerable because you never know, what is really inside a file (just one example, malware) [4].

The three main types of applicable trust after Rosseau [7] are (1) trusting beliefs, (2) trusting intentions, and (3) trusting behaviors. These terms can be defined as follows: 1) Trusting beliefs means a secure conviction that the other party has favorable attributes (such as benevolence, integrity, and competence), strong enough to create trusting intentions. 2) Trusting intentions means a secure, committed willingness to depend upon or to become vulnerable to, the other party in specific ways, strong enough to create trusting behaviors. 3) Trusting behaviors means assured actions that demonstrate that one does in fact depend or rely upon the other party instead of on oneself or on controls. Trusting behavior is the action manifestation of willingness to depend.

Each of these generic trust types can be applied to trust in IT. Trusting behavior-IT means that one securely depends or relies on the technology instead of trying to control the technology [5]. Another point of view is the similarity of trusting people and trusting technology, especially information technology, where the main difference lies in the application of trust in the specific area: "The major difference between trust in people and trust in IT lies in the applicability of specific trusting beliefs. People and technologies have both similar and different attributes, and those similarities and differences define which trusting beliefs apply [6].

With trust in people, one trusts a morally capable and volitional human; with trust in IT, one trusts a human-created artifact with a limited range of behaviors that lack both will and moral agency [7]. Because technology lacks moral agency trust in technology reflects the belief about technology's capability rather than its will or its motives [6].

Trust in IT has several interesting implications. First, trust in IT should influence use or adoption of technology. Unless one trusts a software product to reliably fill one's needs, one would not adopt it. Second, trust in IT is a general assessment of the technology that probably affects other IT perceptions, such as relative advantage or usefulness of the technology. Thus, it may influence beliefs and attitudes that affect intentions to use a technology.

Trust in technology is built the same way as trust in people [5]. In their work "Not so different after all: A cross-discipline view of trust", [6] state, that trust is the willingness to be vulnerable, willingness to rely upon and confident and positive expectations. "However, the compositions of trust are comparable across research and theory confusing on parties both inside and outside firms and investigate trust relations from different disciplinary vantage points."

Rosseau state that Trust is a psychological state comprising the intention to accept vulnerability based upon positive expectations of the intentions or behavior of another [7]. Another very interesting publication about the trust in information sources is given by Hertzum [8]. Trust in information sources: seeking information from people, documents, and virtual agents. They compare the notion of trust between people and virtual agents, based on two empirical studies. The testimonials were software engineers and users of e-commerce systems.

Some relational aspects concerning trust in the industrial marketing and management sector can be found in "Concerning trust and information" from Denize, Young [9].

When we come into trust concerning trusting in data and trusting the sources of data, the term "Data Provenance" takes into account. Data Provenance means the origin and complete processing history of any kind of data. A quite good introduction and overview can be found in "Data provenance – the foundation of data quality", from Buneman [10]. "We use the term data provenance to refer to the process of tracing and recording the origins of data and its movement between databases." and "It is an issue that is certainly broader than computer science, with legal and ethical aspects" [11].

Several problems concerning data provenance are covered in "Research Problems in Data Provenance" from [12] and a very good approach for measuring Trust is given by [13] in their paper "An Approach to Evaluate Data Trustworthiness Based on Data Provenance".

4.3 Security and Trust

IoE has widely distributed applications available for a variety of users. It has applications varying from domestic use over to wearable devices and the advanced use, for example, temperature controller of the boiler in the thermal plant. The trust in most applications is required for users to guarantee the reliability or availability of system services. For example, a distributed data storage application would want to guarantee that data stored by a user will always be available to the user with high probability and that it will persist in the network (even if temporarily offline) with a much higher probability [14]. Privacy along with reliability is needed so that the users whose data has been stored in a distributed system would guarantee to protect the content from being accessed by unauthorized users. There are several solutions to it with one easy solution is to encrypt all data before storing. However, in some applications access to unencrypted data is necessary for processing. It may also be sufficient to separate sensitive data from subject identities, or use legally bound strict privacy policies [15–18]. Anonymity is a specific application of privacy where users may only be willing to participate if a certain amount of anonymity is guaranteed. This may vary from no anonymity requirements, to hiding real-world identity behind a pseudonym, to requiring that an agent's actions be completely disconnected from both his real-world identity and his other actions. A reputation system would be infeasible under the last requirement.

4.4 Trust Models

Security is an important cornerstone for the Internet of Everthing (IoE). More specific, all common aspects of security must be regarded. With the huge amount of data created by IoE, the integrity of data and trust in the services offering the data is crucial. Further, to protect important data and user interests, confidentiality of data and the privacy of users must be ensured. In addition to integrity and confidentiality, each request and response inside the IoT have to be authenticated in a proper and secure way. Trust is the subjective probability by which an individual expects that another individual performs a given action on which its welfare depends [19]. Reputation is the collected and processed information about one person/entity in a community from its former behavior as experienced by others. Figure 4.2 gives an overall view of the high level computational trust engine. The output will compute on the request, selection, decision and evidence. The processes involved are context

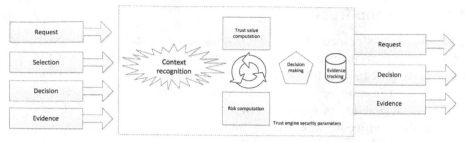

Figure 4.2   High level computational trust engine.

recognition, trust value computation, risk computation which helps in decision making. This process is kept in the record and the engine also has the facility of back tracking which is called the evidence tracking.

According to Eschenauer et al. [25], trust is defined as "a set of relations among entities that participate in a protocol. These relations are based on the evidence generated by the previous interactions of entities within a protocol. In general, if the interactions have been faithful to the protocol, then trust will accumulate between these entities." Trust has also been defined as the degree of belief about the behavior of other entities (or agents) [26], often with an emphasis on context [27].

Figure 4.3 explains how trust (i.e., subjective probability of trust level) and trustworthiness (i.e., objective probability of trust level) can differ. In Figure 4.3, the diagonal dashed line is assumed to be marks of well-founded

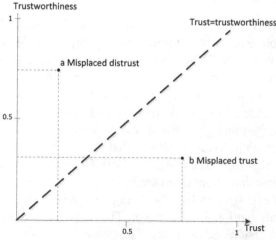

Figure 4.3   Trust level.

trust in which the subjective probability of trust is equivalent to the objective probability [28]. In this Subjective and objective trust are supposed to be interchangeable entities. We have noticed that most of the current trust and reputation models in the literature follow these four general steps:

1. Collecting information about a certain participant in the community by asking other users their opinions or recommendations about that peer.
2. Aggregating all the received information properly and somehow computing a score for every peer in the network.
3. Selecting the most trustworthy or reputable entity in the community providing a certain service and effectively having an interaction with it, assessing a posteriori the satisfaction of the user with the received service.
4. According to the satisfaction obtained, a last step of punishing or rewarding is carried out, adjusting consequently the global trust (or reputation) deposited in the selected service provider [20].

4.4.1 Fuzzy Trust Model Description

An entity's trustworthiness is the quality indicator of the entity's services, which is used to predict the future behavior of the entity (stored in sensors or sensor-embedded things). Intuitively, if it is trustworthy enough, the entity will provide good services for future transactions. In most trust models, the domain of trustworthiness is assumed to be [0,1]. Since the key issue in investigating fuzzy problems is to establish membership functions (membership degrees) by employing the fuzzy set theory, we have to create the mathematical model of fuzzy trust firstly [21].

Suppose that $SN = \{SN_1, SN_2, \ldots, SN_n\}$ is a problem domain of the fuzzy trust model. Here SN_i *(where i= 1, 2,..., n)* is a subset in the corresponding domain. Then we can get the following mapping:

$$MappingFunction : SN \times SN \rightarrow [0,1],$$
$$(SN_i, SN_j) \rightarrow \psi(SN_i, N_j) \in [0,1]. \tag{4.1}$$

where $\psi(SN_i, SN_j)$ represents the degree of trust relationship between SN_i and SN_j. *MappingFunction* is a fuzzy relation mapping from $SN \times SN$ to [0,1].

In this scheme, a neighbor monitoring process is used to collect information of the package forwarding behaviors of the neighbors. Each sensor node in the network maintains a data forwarding transaction table (DFT),

$$DFT = \langle Source, Destination,\ RF_{i,j},\ F_{i,j},\ TTL \rangle \qquad (4.2)$$

where Source is the trust of evaluation evaluating nodes, Destination is the evaluated destination nodes, $RF_{i,j}$ denotes the times of successful transactions which node SN_i has made with node SN_j and $F_{i,j}$ denotes the positive transactions [22].

4.4.2 Reputation Evaluation

Node SN_i evaluates the reputation of node SN_j with which it tries to make transactions by rating each package forwarding process as either positive or negative, depending on whether SN_j has completely done the transaction correctly.

Con denotes the evaluation of the whole metrics in order to judge whether this transaction is successful. The *Con* is computed by

$$Con = [EPFR,\ AEC,\ PDR].[\alpha\ \beta\ \gamma]^T$$
$$= \alpha \cdot EPFR + \beta \cdot AEC + \gamma \cdot PDR \qquad (4.3)$$

where

$$EPDR = \frac{\sum_i^k RECV_i}{\sum_i^k SEND_i},\quad 0 \le k \le n,$$

$$AEC == \frac{\sum_{i+1}^k consume_i}{\sum_i^k send_i + recv_i + \tau} \qquad (4.4)$$

and PDR is packet delivery ratio

where α, β, γ represent the corresponding aspect weights of the different resources. We also define a parameter *SatThreshold* to describe the satisfaction degree. That means, if, then it indicates that node SN_i gets a negative reputation evaluation to node SN_j; if $Con \ge SatThreshold$, it indicates that node SN_i gets a positive reputation evaluation to node SN_j.

The reputation evaluation of all interactions from node SN_i to node SN_j is defined as follows:

$$\delta = \frac{F_{ij}}{RF_{ij}} \in [0,1] \qquad (4.5)$$

Reputation evaluation is the basis of trust management. In our trust model, the reputation is evaluated considering three metrics, *EPFR, AEC* and *PDR*. Compared with other reputation evaluation methods, we consider more factors which can more accurately evaluate the behaviors of nodes according to specific characteristics of IoT [23].

4.4.3 Eigen Trust Algorithm

A natural way to do this in a distributed environment is for peer i to ask its acquaintances about their opinions about other peers. It would make sense to weight their opinions by the trust peer i place in them:

$$t_{ik} = \sum_j c_{ij} c_{jk} \qquad (4.6)$$

where t_{ij} represents the trust that peer i places in peer k based on asking his friends. It can be written in matrix notation: If C is defined as the normalized trust of matrix $[c_{ij}]$ and t_i to be vector containing the values t_{ik}, then $\vec{t_i} = C^T \vec{c_i}$. (Note that $\sum_j t_{ij} = 1$ which is the maximum value of trust of a network). The first thing that we can see when we want to compute the global trust vector t

$$\vec{t_i} = (C^T)^n \vec{c_i}. \qquad (4.7)$$

is that this equation does not depend on i, because for all i the same vector is calculated. Such that c_i can be replaced with any distribution one can think, and because we want when starting all peers have the same chance, we can put the uniform distribution

$$\vec{t} = (C^T)^n \vec{e} \qquad (4.8)$$

where $e_i = 1/m$ for all i, and m is the number of peers in the network.

Secondly, it will not be a good idea to compute the nth power of the matrix C, one reason being the fact that we don't know.

By using the probabilistic interpretation of c_{ij} and looking at the PageRank computation it was found that the following algorithm works perfectly [24].

Simple Eigen Trust Algorithm

$$\vec{t^{(0)}} = \vec{e}, \qquad e_i = \frac{1}{m}$$

repeat

$$\vec{t^{(k+1)}} = C^T \vec{t^{(k)}},$$

$$\partial = \left\| \vec{t^{(k+1)}} - \vec{t^{(k)}} \right\|$$

until $\partial < \in$

There are some big drawbacks with this algorithm. The first is the lack of a prior notion of trust, then newcomers cannot gain trust (the Random-Surfer will get stuck when reaching a peer that has all $c_{ij} = 0$. Also a newcomer will not be known by the other peers). The third issue is that malicious peers can form *groups* or *collectivities* and in such situation the Random-Surfer can get stuck in these collectivities increasing the trust of these peers and decreasing the trust of all other peers. We address in more detail these problems.

4.4.4 Notion of Trust

In the algorithm above, it has been chosen to start with a uniform distribution over all peers, but this strategy doesn't hold if there can be malicious collectives, since if for example, one quarter of the network is forming such a collective, then there is a probability of ¼ to choose them in the beginning, getting stuck later. Also is not good to choose a peer i and start with $\overrightarrow{c_i}$. Thus we need to know a trusty peer to start with. A natural way to choose a peer or better *a set* of trusty peers is to consider the peers who first joined the network. We will call these peers **pre-trusted peers**, though these peers will not have special duties. The only thing that is required is that there are peers in the network that can be trusted, and they will never change their side, becoming malicious peers. The peers who join first (or build) the network have no interest to subvert the network.

Thus if there are P pre-trusted peers, distribution $\overrightarrow{p}$ can be on them by setting:

$$\pi = \begin{cases} \frac{1}{p} & if \; i \in P \\ 0 & otherwise \end{cases} \tag{4.9}$$

4.5 Example: PGP Web of Trust

First, PGP uses a public-key ring: The public-key ring is a structure that store public-key of other users known by the public-key ring user. Essential to this approach is you are assured that a specific certificate is belonging to someone is valid; you can sign the copy of this certificate on your key-ring to guarantee (attest) that the certificate is checked and authentic. If the approval of the certificate should be public, the signature can be placed on a certificate server – thereby making it public.

The public key management of PGP is important. A possible tampering of a public key is a major problem and should be prevented by the key management structure.

In the PGP a web-of-trust is used due to the use in informal environments. The PGP trust model is a cumulative trust model [31]. A certificate can belong from a chain of trust or might be trusted directly. In PGP digital signatures is a kind of introduction. First, when a PGP user signs another's key, the person becomes an introducer of that key. When process grows, it will create a web-of-trust. In another word can any user act as a certifying authority. A PGP user will be able to validate another PGP user's public key certificate [31]. The certificate is only valid to other users if the relying party recognizes the validator as a trusted introducer. In order to use such a system, a level of trust is needed. In the public-key ring there is for each entry a key legitimacy field. This field indicates the extent to which PGP will trust the validity of user's public-key. High level of trust will give a stronger binding of the user ID and the key. The signature trust field indicates the degree to which the PGP user trust the validity of the signer. The last field is the owner trust field. This field indicates the level of trust to which this public-key is trusted to sign other public-key certificates. The three fields are contained in a trust flag byte [32].

When a user places a new public-key in public-key ring, a trust flag byte value must be inserted. If the key is present in the secret-key ring, the key has an ultimate trust is assigned to the trust field. Otherwise, will PGP asks the user which trusts level that should be assigned to the key. The levels can be: Completely trusted, marginally trusted, untrusted or unknown. This is the OWNERTRUST field.

To the public key, one or more certificates can be attached. PGP checks the public-key ring to see if the owner of the public-key is among the known public-key owners. If this is true the SIGTRUST value is set to OWNERTRUST value. If not the SIGTRUST value is set to the unknown user.

The key legitimacy field is calculated by looking at the signature fields in the entry. If just one signature has signature trust value equal to ultimate, then the key legitimacy level is set to complete.

The value of the key legitimacy field is calculated using the signature trust fields found in the entry. If one signature has the signature trust value of ultimate the key legitimacy value is complete. If not then PGP calculates a weighted sum of trust values. Again here some user configurable parameters X and Y are used. $1/X$ is given to a signature which is always trusted and $1/Y$ is given to a signature which is usually trusted. When total weight of the introducers (key – UserId combination) is reaching 1, the signature is trustworthy and the key legitimacy value is set to complete.

As it can be seen PGP is using both personal beliefs and on top of this computational trust.

4.6 Example: X.509 Certificates

As a part of the X.500 recommendations the X.509 standard defines a directory service. The directory service is a server – often more servers in a distributed setup. The directory service has a database containing information about users and map information about users to network addresses.

The X.509 certificate is very important since it is used in IPSec, S/MIME and SSL/TLS [33].

An X.509 certificate contains the public key of the user and is signed with a key from a trusted CA (Certificate Authority) and is based on the use of public key cryptography. The certificate contains information about the CA and the user (the public key of the user). Furthermore, the certificate contains information about algorithms and parameters. The information is signed by calculating a hash value. A digital signature is then created using the hash value and the CA's private key.

A certificate is unforgeable. Therefore, the certificates can be placed in a public directory.

Regarding trust, it is obvious that there will be a common trust to the CA if all are using the same CA [34]. If B sends a message to A the message will be encrypted with A's public key and thereby the message is protected against eavesdropping. And if A signs a message with the private key the message is unforgeable. But in fact, it is not practical that all users are subscribing to the same CA. It is more useful to have several CA because each user must have a copy of the CA's public key to verify the signature and the public key must be delivered to each user in a secure way in order to have trust in the public key. In that way, using more CAs, only a fraction of the users is provided with the public key. The situation is now that users that want to communicate could have two or more different CAs. Therefore, a certificate cannot be verified, unless two or more CAs has exchanged their public keys. From this, in the case of several CAs, it can be concluded that a chain of certificates is necessary. In X.509 is using a hierarchy using forward and reverse certificates. Forward certificates are generated by other CAs. Reverse certificates are generated by X and are certificates of other CAs.

Revocation of certificates is possible in X.509. In the light of trust (a certificate can exceed a period of validity) the certificate must be revoked if the user is no longer certified by the CA. Or the CA's certificate or the user's private key is compromised.

Therefore, a CA must maintain a list of revoked certificates. The user must determine whether a certificate is revoked when receiving a message. Also, a user can maintain a local cache containing a list of revoked certificates.

Belief is the signature. Beliefs also rely on the chain of CAs. It is possible to delete certificates and also decide what a signature can be used for. The computational part is rather binary; either the signature is trusted or not.

4.7 Summary

In this chapter, Internet of Everything and trust have been introduced. Necessity of trust in internet world has been brought up. It also contains the difference between conceptual trust and computational trust. Various computational trust algorithms have been discussed. Finally, Pretty Good Privacy and Certificates have been discussed as an example of computational trust. We expect to see in the future a much-increased use of computational trust, especially in thing-to-thing communication.

References

[1] McKnight, D. H., and Chervany, N. L.; "The Meanings of Trust": Scientific report, 1996 University of Minnesota.
[2] Mayer, R. C., Davis J. H., Schoorman F. D.; "An integrative model of organizational trust". Academy of Management Review. 20 (3), 709–734, 1995.
[3] Bamberger, Walter (2010). "Interpersonal Trust – Attempt of a Definition". Scientific report, Technische Universität München. Retrieved, 2011.
[4] Hörmanseder, R., Jäger, M.; Cloud Security Problems Caused By Virtualization Technology Vulnerabilities And Their Prevention, in: Doucek Petr, Chroust Gerhard, Oskrdal Vaclav (Eds.): IDIMT-2014 Networking Societies – Cooperation and Conflict, Series Schriftenreihe Informatik, Number 43, Page(s) 373–383, Trauner Verlag, 2014.
[5] McKnight 2005; Trust in Information Technology, In G. B. Davis (Ed.), The Blackwell Encyclopedia of Management. Vol. 7, Management Information Systems, Malden, MA: Blackwell, pp. 329–331.

[6] Markus Jager, Josef Kueng; Introducing the Factor Importance to Trust of Sources and Certainty of Data in Knowledge Processing Systems – A new Approach for Incorporation and Processing; Proceedings of the 50th Hawaii International Conference on System Sciences |2017.

[7] Rosseau Rousseau, D. M., Sitkin, S. B., Burt, R. S., and Camerer, C.; Not so different after all: a cross-discipline view of trust, Published in Academy of Management Review, 1998.

[8] Hertzum, M., Andersen, H., Andersen, V., and Hansen, C.; Trust in information sources: seeking information from people, documents, and virtual agents. Elsevier Interacting with Computers 14 (2002) 575–599.

[9] Denize, Young 2007, Concerning trust and information, Published in Elsevier, 2007.

[10] Buneman, Davidson; "Data provenance – the foundation of data quality"', University of Edinburgh, University of Pennsylvania; 2010.

[11] Data Provenance: Some Basic Issues, Published in LNCS: Foundations of Software Technology and Theoretical Computer (FST TCS 2000).

[12] Tan Research Problems in Data Provenance, University of California, Santa Cruz, 2004.

[13] Dai et al., 2008, An Approach to Evaluate Data Trustworthiness Based on Data Provenance, Published in the Proceedings of the 5th VLDB Workshop on Secure Data Management (SDM'08), ACM.

[14] J. Kubiatowicz, D. Bindel, Y. Chen, P. Eaton, D. Geels, R. Gummadi, S. Rhea, H. Weatherspoon, W. Weimer, C. Wells, B. Zhao, OceanStore: An Architecture for Global-scale Persistent Storage, in: Proceedings of ACM ASPLOS, ACM, 2000.

[15] P. Maniatis, M. Roussopoulos, T. Giuli, D. S. H. Rosenthal, M. Baker, Y. Muliadi, Preserving peer replicas by rate-limited sampled voting, in: 19th ACM Symposium on Operating Systems Principles (SOSP 2003), 2003.

[16] Reiter, M. K., Rubin, A. D., Crowds: Anonymity for web transactions, in: ACM Transactions on Information and System Security, 1998.

[17] G. Agarwal, M. Bawa, P. Ganesan, H. Garcia-Molina, K. Kenthapadi, N. Mishra, R. Motwani, U. Srivastava, D. Thomas, J. Widom, Y. Xu, Vision Paper: Enabling Privacy for the Paranoids, in: VLDB, 2004.

[18] G. Agarwal, M. Bawa, P. Ganesan, H. Garcia-Molina, K. Kenthapadi, R. Motwani, U. Srivastava, D. Thomas, Y. Xu, Two Can Keep a Secret: A Distributed Architecture for Secure Database Services, in: CIDR, 2005.

[19] Gambetta, T.: Can we trust trust? In: D. Gambetta (Ed.), Trust: making and Breaking Cooperative Relations, Basil Blackwell, Oxford, 213–238. (1990)

[20] Sergio Marti, Hector Garcia-Molina, Taxonomy of Trust: Categorizing P2P Reputation Systems, Stanford University {smarti, hector} @cs.stanford.ed

[21] Azzedin, F., Ridha, A., Rizvi, A.: Fuzzy trust for peer-to-peer based systems. In Proc. of World Academy of Science, Engineering and Technology, Vol. 21, 123–127. (2007)

[22] Dong Chen, Guiran Chang, Dawei Sun, Jiajia Li, Jie Jia, and Xingwei Wang, TRM-IoT: A Trust Management Model Based on Fuzzy Reputation for Internet of Things, ComSIS Vol. 8, No. 4, Special Issue, October 2011.

[23] David Challener, Kent Yoder, Ryan Catherman, David Safford, Leendert Van Doorn, A Practical Guide to Trusted Computing.

[24] Adrian Alexa, Reputation Management in P2P Networks: The EigenTrust Algorithm.

[25] L. Eschenauer, V. D. Gligor, and J. Baras, "On Trust Establishment in Mobile Ad Hoc Networks," Proc. 10th Int'l Security Protocols Workshop, Cambridge, U.K., Apr. 2002, Vol. 2845, pp. 47–66.

[26] L. Capra, "Toward a Human Trust Model for Mobile Ad-hoc Networks," Proc. 2nd UK-UbiNet Workshop, 5–7 May 2004, Cambridge University, Cambridge, UK.

[27] H. Li and M. Singhal, "Trust Management in Distributed Systems," Computers, Vol. 40, No. 2, Feb. 2007, pp. 45–53.

[28] Solhaug, D. Elgesem, and K. Stolen, "Why Trust is not proportional to Risk?" Proc. 2nd Int'l Conf. on Availability, Reliability, and Security (ARES'07), 10–13 April 2007, Vienna, Austria, pp. 11–18.

[29] Christer Andersson, Jan Camenisch, Stephen Crane, Simone Fischer-Hübner, Ronald Leenes, Siani Pearson, John Sören Pettersson, and Dieter Sommer, Karlstad University, Sweden; Trust in PRIME, 2005 IEEE International Symposium on Signal Processing and Information Technology.

[30] John Sören Petterson, Simone Fischer-Hübner, Jenny Nilsson, Ninni Danielsson, Mike Bergmann, Thomas Kriegelstein, Sebastian Clauß, Henry Krasemann: Making PRIME Usable. In Proceedings of the Symposium of Usable Privacy and Security (SOUPS), 4–6 June 2005, Carnegie Mellon University, Pittsburgh, PA, USA. July 2005.

[31] http://www.pgpi.org/doc/pgpintro/#p12

[32] Network Security Essentials Application and Standards 6th ed. W. Stallings, Pearson 2017, appendix 4, p. 15.

[33] Network Security Essentials Application and Standards 6th ed. W. Stallings, Pearson 2017, p. 139.

[34] Network Security Essentials Application and Standards 6th ed. W. Stallings, Pearson 2017, p. 142.

5

Security in Internet of Things

Egon Kidmose and Jens Myrup Pedersen

5.1 Introduction

2016 was a year when the discussions about Internet of Things and security gained significant grounds. Not only was it yet another year where the challenges of cybercrime became visible to the general public, maybe with the presumable Russian hacking of Hillary Clinton's emails as the most prominent example, but at the end of the year the Mirai Botnet used Internet of Things devices to perform successful attacks on several Internet infrastructure points.

On October 21, the DNS infrastructure provider DYN suffered a DDoS attack originating from the Mirai botnet [1], with an expected 100,000 endpoints/devices being used for the attack. As a consequence a number of well-known websites were down, including Spotify, Twitter and Github. The attack was rather advanced in the sense that it used DNS water torture, where recursive DNS retry traffic is amplifying the attack and increasing its impact further. In the aftermath, it seems that only part of the Mirai botnet took part in the attack, and that also other devices were involved. According to DYN the attack took place during two waves, the first lasting around two hours and the second around one hour. The attack was reported to reach 1.2Tbit/s.

It was only a few days until the Mirai botnet would again take the headlines of news media worldwide, with an attack on the Liberian Internet infrastructure on November 3. Forbes wrote "Someone Just Used The Mirai Botnet To Knock An Entire Country Offline" [2], but afterwards it turned out that probably this was a bit of an exaggeration as there were no reported national-wide Internet outage. However, this does not mean that the attack was without serious implications.

One of the persons who has been intensively following the Mirai botnet attacks is Brian Krebs, who runs the security blog "Krebs on Security" [3]. He fell victim himself to one of the earliest known attacks by the network, which took place on September 20.

As we shall also discuss later in the chapter, the Mirai botnet is interesting not only because of its size, and the attacks it was able to perform, but also because of the underlying weaknesses in the devices used. According to Symantec the botnet mainly consists of routers, digital video recorders, surveillance cameras and other kinds of smart Internet-connected appliances [4]. Based on the source code, which was released, it is expected that the Mirai botnet managed to take control of an estimated 500,000 devices based on only 62 combinations of default usernames and passwords. Due to the nature of the botnet, where infected devices scan their surroundings in search for other potential victims, Symantec estimates that such devices are on average scanned every 2 minutes.

The Mirai botnet might very well be just the beginning. We are seeing more and more devices being connected to the Internet, not only in our homes, but also in work places and public spaces. While this gives many useful (and not so useful) applications, it does not come without challenges, not at least in the security domain. First of all, the devices are often small and with limited capabilities, so many will never be patched, and even new devices are often sold with known security flaws and vulnerabilities [5]. Second, user friendliness and plug-and-play is often prioritized higher than security, and most devices comes with either weak or no security by default: For example, a series of devices might all come with the same username and password, and/or there might be no protection against brute force password attacks. Another aspect of the plug-and-play is that many devices are accessible from the global Internet, which in combination with vulnerabilities and poor password security makes it an easy target for attackers.

This chapter will present a number of examples of problematic or compromised IoT devices, and illustrate the trade-offs that often are when designing such devices. Making devices secure, user friendly, and cheap at the same time is not trivial. Providing heavy security frameworks is beyond the scope of the chapter, and these are anyway hard to use in the real world: What is useful and important is to consider security throughout the process of designing, implementing, deploying, and/or configuring systems that involve IoT devices, and make deliberate choices when making tradeoff to e.g. cost and usability. The chapter aims at providing some qualified inspiration and inputs when making such decisions.

5.2 Examples of Problematic IoT Devices

To get an idea of the current situation of security and IoT we will look into a number of recent events, where IoT devices were compromised. In the previous section the Mirai botnet and attacks were described, but in fact the Distributed Denial of Service attack is just one way to abuse the devices. While a powerful tool, it is also a tool that is not targeting the owners/operators of the devices themselves, and it is important to show that other types of attacks exist where for example devices are left failing or misbehaving, or where attackers can steal different kinds of information.

5.2.1 IP Camera

This case is not an attack, but rather a study conducted by students from Aalborg University [6] who conducted a security test on an IP-camera, which is a relatively new device from a reputable manufacturer. In the study, several weaknesses were discovered:

- By default, the camera worked from a fixed IP address with username "admin" and password "admin".
- The camera includes a webserver, hosting a webpage which can be used to configure the camera. An app can be installed which makes it possible to follow the camera from a phone on the local Wifi, and it is also possible to register the camera on a website, which allows for following it from any Internet-connected device.
- When the app is communicating with the camera both usernames, passwords and images are transmitted in cleartext.
- By analysing the network traffic in wireshark it was possible to identify the HTTP command that would request the picture, and even from other devices (e.g. a PC) on the local network the image could now be requested and shown without any authorization.
- In order to do video streaming from the camera authentication is needed, but there appears to be a security flaw since it accepts an empty username and password. A wrong username and password is rejected though.

What was really surprising in the study was the security flaws when allowing connections over the Internet. When connections through the cloud service were allowed, which is the default setting, the following could be observed:

- Even if the default username and password ("admin" and "admin") were changed the new usernames and passwords would be transmitted to the associated cloud service in cleartext over the Internet – in fact, together with the camera ID number, as well as the version number of the software. Furthermore, this information is transmitted to the camera every 30 seconds.

- Creating an account on the website can only be done using HTTP (not HTTPS), and all you need to register is the username, password and device ID number.

5.2.2 Internet Gateways

While it can be discussed whether Internet gateways fall within the IoT defini-tion, they are in any case an important component in the IoT infrastructure, and many of the same challenges as for IoT devices are also present here. We saw in the previous example the implication of poor management of usernames and passwords. This reflects the choices the manufacturers have of practically just three different approaches:

- No security by default. This is by far the easiest (and cheapest) setting to start out with, and users immediately can start using their devices. There is not setup required, and no manuals to read.
- The second option is to enable some security settings by default, e.g. access control and encryption, but rely on default usernames and pass-words, which the user would be encouraged to (but not required to) change before starting to use the device. This is still easy to handle for the manufacturer, but the user will need to learn the default usernames and passwords before starting to use the devices.
- The third option is the more secure approach of providing each produced device with a unique username and password. While this is much preferred over the two other approaches, it increases the complexity in the production process, and it requires the manufacturer to provide individual instructions to each customer, with all the inherent risks that the user forgets or loses the password. One way of conveying this information to the user is by printing it on the device itself, but of course this is only a viable solution in case the device itself cannot be accessed by people who should not have this information.

One company that has used this approach is TP-Link, who also got quite some attention from the security community when in late 2015 Mark C. [7] published the findings that the unique password for Wifi on their Internet gateways were the last 8 digits of the MAC address (which is broadcasted all the time). This demonstrates nicely how important it is not only to provide unique usernames/passwords, but also to ensure that they cannot be derived from available information.

Another challenge with the Internet gateways is the use of the Universal Plug and Play (UPnP) protocol, which often implement the Internet Gateway Device (IGD) protocol that allows devices otherwise limited to internal networks to be accessed from the global Internet. Many residential homes are somewhat protected from outside threats due to the use of Network Address Translation (NAT), which at least makes it harder to access devices on an internal network. In short, devices on the internal network only has internal addresses, and can only be reached from the outside if the initial connection is established from the device itself or if the NAT gateway is configured to forward requests on specific ports to that device. However, this is exactly what UPnP/IGD is doing. Therefore, there is a high risk that the unaware user will not know that devices on the internal network now can be directly accessed from the global Internet. On the other hand this is still better than when users due to lack of knowledge set up a broad spectrum of port forwardings (so he doesn't have to mess with it next time something does not work as expected).

5.2.3 Smart Energy Meters

Smart meters are a broad term for devices that monitors energy usage, such as heat and electricity, sometimes extending into also other utilities such as water supply, and transmit this information to e.g. utility companies who bills the customer accordingly. While new electronic meters are initially mainly used for billing purposes, it is expected that in the future the usage will actually become smarter, and used for e.g. for controlling the energy usage either by the users or automatically through e.g. demand/response systems. Therefore, it is expected that the meters will play a vital role in both energy systems and systems for smart homes.

When IoT is moving into the energy system, it is also moving into some of the most critical infrastructures a country can have. Being able to take a country (or some specific targets) off the grid can seriously cripple a country. This was seen with Ukrain's power grid in late 2015, when hackers managed to take control over three power distribution centers, and locked out the local operators who could just helplessly watch as the mouse curser moved over the screens and took out the power distribution to 230,000 residents – as well as the backup power supplies to two of the distribution centers themselves [8]. In the following analysis of the attacks, it seems that the first part of the attack was quite classical: A spear-fishing attack against the IT staff and system administrators working within the electricity distribution in Ukraine was used to install a program called BlackEnergy3 to the machines, opening a backdoor for the hackers (exploiting the macro feature in Microsoft Word). This is still far from the SCADA networks that were critical, but with time and effort they managed to use the access to harvest worker credentials for the VPNs used to remotely enter the SCADA network. From there it went on more sophisticated (and very targeted), with malicious firmware for devices in the grid, which would prevent operators from easily taking control of the system again.

In 2014 independent researchers Javier Vidal and Alberto Illera analyzed the security in Spanish smart meters [9]. They took some of the smart meters apart – physically – and inside they found the encryption keys and unique identifiers, which should ensure that the nodes higher in the system know whom they communicate with, and that no one can tamper or access this information. With this information handy, it would be possible to either submit wrong information, for example to lower the electricity bills, or to submit information using wrong ID numbers. While the consequences of such an attack do not immediately appear comparable at all with what happened in Ukraine, we could expect that over time the data from the meters will be used for much more than just accountings, such as providing real-time inputs for the power production, or with a capability of controlling power consumption of smart home devices. This point illustrates another challenge with security and IoT: Even if devices are designed with security in mind, this would often reflect the threat picture at the time of production, which would again reflect the intended use of the device as well as the motivations and capabilities of potential attackers. While these things change over time, the devices most often do not unless specifically designed for.

The analysis of Spanish smart meters does not stand alone. Denmark has been one of the first-movers with respect to integration of sustainable energy in the power systems, and the smart meters are expected to play an important role in balancing demands and supplies. While only little is published about it, a Danish student at the Technical University of Denmark managed to obtain access to smart meters of the Danish company SEAS-NVE who recently installed 400,000 of these [10].

In fact, IoT has a much bigger potential in the energy sector than putting smart devices in the home. There is a big potential in installing sensors in power systems, where it can constantly monitor the state of systems and support operations and maintenance, as well as in all other utilities. However, it is important to be aware of what can happen not only if a single device is compromised, but if whole systems or large numbers of devices are taken over and abused. While no physical damage was done in the attack on the power system of Ukraine, attackers have in other cases been able to inflict physical damage from attack on control systems, such as when a German steel plant was attacked [11].

5.2.4 Automotive IoT

The automotive industry is another fast-changing industry, where IoT is expecting to play an increasing role for both operations and user experience. As with many developments, it did not happen from one day to another, but was more gradually introduced: GPS systems are increasingly based on apps like Google maps, the entertainment systems are integrated with streaming services like Spotify, and the trend is towards more integration between car operations and mobile devices. Porsche has for example made an app for the Apple Watch, which allows the user to check the status of the car and even control some elements such as locking the car or folding in the external mirrors.

The idea of using smartphone apps for monitoring or controlling elements of the car seem to be a trend across the automotive industry: VW has for example developed App Connect, Skoda has SmartLink, and Ford has developed Ford SYNC. But maybe the most interesting developments come from the new players in the car industry such as Tesla and the cars we can expect to be launched in the near future from Google and Apple.

Controlling a car – or any other device – from a smartphone initially raises the question whether the phone itself could be remotely controlled by attackers. While there is no real example of this happening today, it is important to be aware of the risks that lay in the combination of increased complexity and resources in the devices together with the devices being more attractive goals for hackers as the devices are used for critical things such as car control.

As long as separate systems are kept separate, the damage should be limited. After all, if all a hacker can control is what music is being played there is not much damage done. This seems quite optimistic though, as was demonstrated by Charlie Miller and Chris Valasek who in 2015 demonstrated how they could hack a Jeep Cherokee. They managed to get remote access not only to the cars entertainment system, but through the entertainment system they were allowed to send commands to dashboard functions, steering, brakes and transmission [12]. While this is bad enough in itself, it of course also allows hackers to monitor everything that is going on in the car and with the driver, which opens op many potential breaches of privacy. Another interesting aspect of this story is that the patch released by Chrysler cannot be downloaded or installed automatically, but requires either a mechanic or installation through a USB stick. While this might be a viable approach for a one-off update for a car, it does not seem to be a viable approach for keeping all IoT devices and sensors patched.

A year later, Charlie Miller and Chris Valasek were back and demonstrated how far they could go if allowed to plug in a laptop directly to the Jeep's CAN network through a port under the dashboard. This made it possible to e.g. disable breaks at any speed, turning the steering wheel at any speed, and causing unintended acceleration [13]. This immediately demonstrates the potential danger if it becomes possible to access these systems from the outside, e.g. through the often less secured entertainment systems. But it also raises the question of what physical access can do: Imagine for example what a mechanic (or any other person with malicious intent) can do to the car if physical access can be achieved.

While Miller and Valasek are probably the most famous "car hackers" they are not alone. In June 2016, Ken Munro managed to hack a Mitsubishi Outlander hybrid, leading Mitsubishi to recall at least 100,000 vehicles [14]. The hack is more detailed described in [15]. Munro took advantage of the fact that the Mitsubishi was controlled through Wifi instead of GSM, requiring the user to make use of the Wifi hotspot offered by the car in order to control it. While the WiFi is in fact secured with a unique pre-shared key (which is communicated to the user on a piece of paper included in the owner's manual) the length and format made it possible to simply crack it. Unfortunately, it is not possible to change the pre-shared key. The SSID comes in the specific format "REMOTEnnaaaa", where the "n"s are numbers and the "a"s lowercase letters. An interesting feature of this is that using services that map out Wifi networks such as wigle.net it is rather easy to get a map of all these cars (or even find the specific one you are looking for). Getting back to the hack, it is interesting to learn what the hackers were able to do based on the SSID and pre-shared key. By sniffing up the messages they were able to figure out the messaging protocol, and use it for turning lights on and off, forcing the car to charge, play with heating and air conditioning and finally to disable the theft alarm.

In the future, we are likely to see more sensors connected wirelessly around the car, measuring everything from temperature in different places to oil levels and tire pressure, but also driving conditions with sensors and cameras observing the surroundings. While each of these separately seems innocent, it is important to be aware of the consequences if one or multiple sensors are hacked and shows wrong information – especially if the information is used for making the car take automated action, e.g. breaking and steering (or not breaking and steering), for example in self-driving cars.

5.2.5 IoT and Health

IoT also has great potentials in the health and care sectors. For a start, the focus has been on development of devices and systems for monitoring of e.g. heat activity, respiratory rate, skin temperature, blood pressure, oxygenation and much more [16]. This information is then manually or automatically processed, and used for generating e.g. alarms. We have already seen steps being taken into the next level, where not only sensors but also actuators are embedded in patients, making it possible to e.g. adjust medication based on the information collected by sensors.

It goes without saying that it is can be critical if hackers access and alter with these kinds of systems, and as in the previous cases it is obvious that they can be used both as weapons and attack points. For the monitoring part, it seems fairly harmless if just a single device is attacked, but if many devices (or the whole system) is taken down this could pose a real threat, and at least make the users feel unsafe. For the actuators, it all depends on what the actuators can do, and/or the consequences of them not doing what they were supposed to do.

While there are apparently few examples of attacks on IoT health devices yet, in principle all of the concerns raised in the previous sections are still valid here. And there is every reason to be aware of the threats against the health sector, where the severe consequences of attacks make ideal victims for e.g. ransomware and other kinds of blackmailing. During some of the ransomware peaks in 2016, e.g. the Locky Ransomware campaigns, health care was the most affected industry, followed by telecom, transportation and manufacturing [17]. Other companies are publishing similar numbers for other ransomware families, e.g. the Cryptowall [18]. It is unknown how many hospitals actually end up paying the ransom, but it is probably fair to assume that some will end up paying, considering that the consequences can otherwise be fatal. Along the same line, it is probably also fair to assume that not all attacks are publicly known or even counted in the various statistics published.

One example of a hack against a well-known health-oriented IoT device is the Fitbit-hack, where Axelle Apvrille demonstrated that she was able to send infected packets to a Fitbit-tracker, potentially allowing for infections on any computer connected to the tracker if an appropriate exploit exists. She also demonstrated that she could manipulate the activities logged by the tracker [19]. After the hack was published, Fitbit entered the discussion and disputed the claim [20].

Security weaknesses have also been found in more "serious" IoT health device systems, some of which include active devices which can severely impact a person's health if used with malicious intent. One example is the Merlin@home transmitter, which is used for forwarding data between implantable cardiac devices and hospitals/physicians. It can be used both for transmitting measurements from the patients to the hospitals/physicians, but also on the other hand to send commands to the device, e.g. in order to deliver paces or shocks. On January 9 2017 the U.S Food and drug administration sent out a warning of a weakness discovered, including the following problem summary [21]:

> "Many medical devices – including St. Jude Medical's implantable cardiac devices – contain configurable embedded computer systems that can be vulnerable to cybersecurity intrusions and exploits. As medical devices become increasingly interconnected via the Internet, hospital networks, other medical devices, and smartphones, there is an increased risk of exploitation of cybersecurity vulnerabilities, some of which could affect how a medical device operates.

> The FDA has reviewed information concerning potential cybersecurity vulnerabilities associated with St. Jude Medical's Merlin@home Transmitter and has confirmed that these vulnerabilities, if exploited, could allow an unauthorized user, i.e., someone other than the patient's physician, to remotely access a patient's RF-enabled implanted cardiac device by altering the Merlin@home Transmitter. The altered Merlin@home Transmitter could then be used to modify programming commands to the implanted device, which could result in rapid battery depletion and/or administration of inappropriate pacing or shocks."

This is interesting not only because it demonstrates that devices with such weaknesses can find their way to the market, but also because it is an example of increasing regulatory attention to the challenges. The future could very well bring in closer monitoring and regulation of the market for IoT devices, similar to what we already know from administration of medicine.

5.2.6 The Smart Home and Appliances

Most of the devices discussed in the previous sections are somehow related to existing infrastructures, and in different ways distributed through corporate suppliers who have or should have both interests in keeping systems secure, and resources to ensure that it happens.

When it comes to devices for the so-called Smart Homes or similar appliances, often the devices are bought directly by the customers who can range from individual households to large corporate customers. Thus, the customer might or might not have any knowledge or concerns about security issues, and in most countries, there is no legislation in place dealing with consequences of flawed security: This is by and large a market, where customers can go ahead and buy some random cheap devices with little interest and ability to check if they are secure, and even if the product in itself is secure, the customer will often not have the knowledge to install, operate and maintain the devices with security in mind. Some countries have started to pay more attention to the issue, and for also giving developers, manufacturers and service providers legal responsibilities for securing their devices.

The devices, including cameras, mentioned in the introductory discussion about the Mirai botnet are examples of such devices, but in this case the devices attacks were used as weapons towards other services rather than for malicious activities targeting their owners. The list of existing and potential devices for the smart home is almost without limit: Smoke detectors, home surveillance, thermostats, light control, music and entertainment, smart electricity switches, hot water kettles, cleaning robots and smart locks are just some examples. In order to obtain the smartness, these devices should of course communicate and be able to work together.

The list of devices with known security problems here is almost endless, fueled by the fact that it is easy to produce and sell such fascinating and fancy devices, and by a market where price de facto is more important than security (even if people indicate otherwise when interviewed).

One of the appliances that has gotten quite some attention is the smart door locks, which can be operated either via the Internet or simply through Bluetooth. Anthony Rose presented on Def:con in 2016 a practical study of 16 different door locks, of which 12 where hackable in different ways. Some would simply transfer the user passwords in plaintext over a Bluetooth connection, meaning that it is very easy for anyone within Bluetooth range to sniff the password and afterwards operate the door lock. In another case he was able to easily change the password, being able to not only operate the lock but also to lock out the legitimate user (so to speak). And in other cases again, classical implementation flaws such as the use of non-random nonces allowed for gaining control. In line with previous reports on the same problems, one of the locks also worked as fail-open, meaning that failure of the lock would make it open: In the case here the fail was obtained by sending non-standard commands to the lock, whereas in other cases it has been obtained by creating a small DoS-attack by sending lots of requests from a single smartphone. In any case, fail-open might not be the most appropriate design for a door lock [22].

The consequences of vulnerable smart locks were demonstrated as late as January 2017, where the Austrian hotel Seehotel Jägerwirt experiences an untraditional ransom attack. The hackers managed to obtain control of the Hotel's smart locks, and demanded a ransom to be paid in exchange for giving access back to the hotel management. The management ended up paying 1,500 Euros in bitcoin to the attackers. The attack was done just at the beginning of the skiing season where the hotel was fully booked, and came on top of a previous attack where the hotel ended up paying a ransom for regaining access to their reservation system. Interesting enough, it is reported that the hotel plans to go back to old fashioned mechanical keys during the next remodeling [23].

The given examples of door locks illustrate two different scenarios, one where the attack is done over the Internet, and one where the attack requires the hacker to be physically closer to the device attacked (at least at some point in time, e.g. for sniffing the password). Other attacks require physical access in order to compromise the device, which might not be feasible in all environments. One example of this was published on the Black Hat USA conference in 2014, where researchers discovered a vulnerability in the Smart Nest Thermostat which would make it possible to bypass the firmware verification, change the behavior of the device, and use it as a beachhead to attack other nodes within the local network. The attacker could maintain the control and access of the device even after the physical contact was interrupted [24].

Another study from 2016 deals with the overall platform called Smart-Things rather than individual devices. Even if the study got most headlines on their findings with respect to smart locks (they demonstrated an app that would monitor when the PIN code was changed, and send the new PIN code to an attacker, pretending that it was simply a battery level monitor) the study is interesting because it demonstrates security flaws in the underlying platform which provides apps within the framework access to control of the devices – and apparently too much access. Among other things, the researchers were able to inject erroneous events to trick devices, so they for example could set off a fire alarm and flip an electricity switch to vacation mode [25].

5.3 Security Challenges in IoT

In the previous section, we looked at some known examples of how vulnerable IoT devices were exploited. We saw how attackers in many cases have quite easy access to take over the control of IoT devices in order to use them in e.g. Distributed Denial of Service attacks, for information theft, or simply in order to destroy the usual operations.

IoT is still in a very early stage, and the expected increase in the number of IoT devices will just add to the problem, as will the platforms that connect the devices and the users, potentially through third party platforms, applications, and cloud services. It is hard to guess how many devices there will be, and also nontrivial to measure the number – already today there are different estimations of how many devices exist: In 2016 Gartner estimated there to be 6.4 billion devices not including smartphones, tablets and computers [26], whereas IHS Markit estimates the number to be around 9 billion and IHS estimates 17.6 billion including smartphones, tablets and computers [27]. For the future expectations, IHS Markit predicts 30.7 billion devices (including all devices) for 2020, Gartner 20.8 billion, and IDC 28.1 billion (both of the latter excluding smartphones, tablets and computers).

It remains unknown how many IoT devices are actually being hacked, and with the decentralized structure of the Internet and home networks, it is inevitable that there will be large shadow numbers. Also, we may assume that many attacks will never be reported, and even the owners and users of devices may never know that they have been under attack.

When talking about numbers, Shodan [28] is providing an interesting service by constantly monitoring what devices can be seen "from" the Internet. This makes it possible to find interesting statistics, but also to look for devices which should not be seen from the outside. Based on the footprint received, it also provides valuable information of e.g. software versions and potential security flaws.

While the numbers might be one interesting aspect of the security challenges with IoT, it is certainly not the only one: Even more importantly is how critical the usage is. There is a clear trend that both monitoring devices and actuators are being used in increasingly critical environments, including some of the industries discussed in the sections above: The health sector, the automotive sector, and the energy sector.

But maybe most important is the combination of numbers and criticality because it allows for serious attacks on system level rather than device level, for example:

- When a single electricity meter or the electricity control of a single house is attacked, it can have severe consequences for that household – in the worst case, turning on some devices could lead to destruction of the device, or even fires. However, if many households were attacked at the same time it could lead to severe power outages and potentially threaten whole power systems including physical damages that would take long times to recover.
- When traffic control systems are connected and smart cars become integrated, the potential damage of infiltration of the systems also grows: In the worst case a systematic simultaneous attack on cars and road infrastructure could cripple the whole traffic infrastructure and cause severe damage to a city or even country.
- As long as systems are separated, it does not create too much harm if cheap entertainment devices are taken over by attackers – unless they are used as weapons against others in e.g. botnets such as in the Mirai case. But when systems are interconnected and used through vulnerable platforms, it becomes potentially much more dangerous, because there is no "secure zone". There have already been such examples in the automotive industry, and our homes is another zone where many different kinds of devices are connected through the same local networks and thus within easy reach of each other.

With this in mind, some of the most important security challenges within IoT can be listed:

How do we ensure that devices and system platforms are designed and implemented to be secure? This is particularly challenging without any security standards, and with the broad spectrum of users who have little knowledge or concern about the security of the devices. While it is not trivial, it is worth considering if some kind of certification of devices/platforms should be required, as it is already the case for electrical equipment in many countries.

How do we ensure that devices are properly configured and maintained? Many users, especially in small companies and among private consumers, have little knowledge about system configuration and setup. While user awareness is important, there is also a need for user friendly technical solutions. These might be highly automatized, but there is a dilemma in having highly automatized solutions and at the same time giving the user control of his security configurations. Also, we probably need to accept the fact that many devices come "as is" and cannot be patched even if vulnerabilities are discovered.

How do we ensure that vulnerabilities are discovered and countermeasures taken? In the world of computers, many security measures have been implemented at device level, e.g. in terms of anti-virus systems, and even if not perfect these have proven to be rather efficient. For IoT devices this does not immediately seem to be a viable solution due to limited computational capabilities, limited possibilities to update the devices, and a large variety of small tailor made operating systems. Monitoring network activity for suspicious/malicious activity could be one way forward, potentially even with the ability to block malicious behavior thus providing protection even for unpatched devices.

How do we protect the data? IoT is all about collecting and using data, and both for functional and privacy reasons it is necessary to ensure confidentiality, integrity and availability. Breaking data integrity can be used to impact systems, and breaking the confidentiality can be used and abused for malicious and criminal purposes. This challenge seems rather overwhelming as the data from many different sources are collected and combined, making it almost impossible to understand the data flow, let alone controlling it.

5.4 Security Recommendations

As we have seen in the previous part of the chapter, Internet of Things security is not trivial, and security is part of a tradeoff also with functionality, usability and cost. This is challenging in itself, but even more so when the future application of systems is not fully known: A design that provides the user with some protection against theft of information about his energy usage might be OK for that purpose (after all these data are not so interesting), but if access to the data from such devices in the future can be used for launching sophisticated systematic attacks on energy systems, the same design might be largely inefficient.

Working systematically with security in all phases of the devices is highly recommended, and there are already several frameworks out there to be considered. Cisco has proposed a general framework [29], and the Industrial Internet Consortium has proposed another framework specifically for the Industrial Internet of Things [30]. A rather general set of recommendations are provided by the U.S. Department of Homeland Security, who has addressed the challenges by developing a set of Strategic Principles for Securing the Internet of Things [31]. These principles are specifically designed for a large range of stakeholders, i.e. IoT developers, IoT manufacturers, Service providers, and Industrial and business-level consumers, and it specifies recommended roles of each. It is based on six non-binding principles, which are elaborated in a number of suggested practices:

- Incorporate security at the design phase.
- Advance security updates and vulnerability management.
- Build on proven security practices.
- Prioritize security measures according to potential impact.
- Promote transparency across IoT.
- Connect Carefully and Deliberately.

While the frameworks and strategic principles provides good and useful guidelines, they also seem very much like a first step towards a more secure IoT and we yet have to see how they will play out in practice.

For a beginning, a simple recommendation could be to get the basics rights and not repeat the mistakes from the early days of the Internet:

- When designing, deploying, configuring and using a device play the devil's advocate and consider how it could be abused, and take appropriate countermeasures.
- As a user, consider security as a feature before buying and deploying IoT devices. Consider also the tradeoff between functionality, usability, cost and security.
- Use unique usernames and passwords, which cannot be derived from information available for attackers. The passwords should be strong, and there should be protection against brute force attacks e.g. blocking repeated guessing attempts.
- Use strong encryption for all communication.
- By default, and throughout the life of the devices, disable services and ports which are not being used.
- Consider how the product is going to be maintained, and keep IoT devices updated as much as possible.
- Block for Internet access unless specifically required. Make sure to understand how the device is communicating, and that the communication channel is secure.

- Consider limiting the physical access to the device.
- Consider using separate networks for different applications, in order to make sure that critical services are running in secure zones.
- Consider an appropriate level of network based security, such as firewalls and network monitoring systems (including Intrusion detection/prevention systems).
- Consider at both device and system levels continuously the potential threats and their consequences, relevant prevention mechanisms, relevant detection mechanisms, and relevant plans for mitigating and handling attacks should they happen.
- Ensure that all critical data are properly protected, e.g. using encryption.
- Ensure that all critical data and systems are properly backed up.

From a pragmatic point of view, we probably also need to accept that the Internet of Things in the foreseeable future will consist of a good mix of secure and non-secure devices, and that both devices and systems will be compromised. Therefore, building walls and securing critical services and devices through segmentation also seems unavoidable at this stage, even if it comes with a higher price tag.

5.5 Conclusion

There is no doubt that we will see the number of devices connected to the Internet increase, a development that has many positive aspects, and which can help us to develop more comfortable and sustainable ways of living. However, the development also calls for immediate action on the deplorable security practices seen today – otherwise what could be a great opportunity can turn into a significant threat against individuals, organizations and societies. While not attempting to provide complete solutions, hopefully this chapter can inspire both manufacturers, developers and users to carefully consider threats and relevant security measures when planning their IoT strategies and deployments.

References

[1] http://dyn.com/blog/dyn-analysis-summary-of-friday-october-21-attack
[2] http://www.forbes.com/sites/leemathews/2016/11/03/someone-just-used-the-mirai-botnet-to-knock-an-entire-country-offline
[3] https://krebsonsecurity.com

[4] https://www.symantec.com/connect/blogs/mirai-what-you-need-know-about-botnet-behind-recent-major-ddos-attacks

[5] M. Hastings, J. Fried, N. Heninger, "Weak Keys Remain Widespread in Network Devices", Proceedings of the 2016 ACM on Internet Measurement Conference, ACM, 2016.)

[6] J. Frandsen, T. K. Johansen, "Sikkerthed i IoT? Et IP-kameras bekendelser", student report, Aalborg University, 2016.

[7] https://twitter.com/LargeCardinal

[8] https://www.wired.com/2016/03/inside-cunning-unprecedented-hack-ukraines-power-grid

[9] http://www.bbc.com/news/technology-29643276

[10] http://www.dtu.dk/nyheder/2017/01/dtuavisen_hyret-som-hacker?id=af88ae0e-e2c2-4c6d-9f9a-c785fa08fa42

[11] https://www.wired.com/2015/01/german-steel-mill-hack-destruction

[12] https://www.wired.com/2015/07/hackers-remotely-kill-jeep-highway

[13] https://www.wired.com/2016/08/jeep-hackers-return-high-speed-steering-acceleration-hacks

[14] https://www.wired.com/2015/07/hackers-remotely-kill-jeep-highway

[15] https://www.pentestpartners.com/blog/hacking-the-mitsubishi-outlander-phev-hybrid-suv

[16] M. Hassanalieragh, A. Page, T. Soyata, G. Sharma, M. Aktas, G. Mateos, B. Kantarci, S. Andreescu, "Health Monitoring and Management Using Internet-of-Things (IoT) Sending with Cloud-based Processing: Opportunities and Challenges", Proceedings of 2015 IEEE International Conference on Services Computing, IEEE Computer Society, 2015.

[17] https://www.fireeye.com/blog/threat-research/2016/08/locky_ransomwaredis.html

[18] http://www.beckershospitalreview.com/healthcare-information-technology/hospitals-are-hit-with-88-of-all-ransomware-attacks.html

[19] http://www.theregister.co.uk/2015/10/21/fitbit_hack

[20] http://www.forbes.com/sites/bradmoon/2015/10/21/fitbit-trackers-can-be-hacked-infect-pcs

[21] http://www.fda.gov/MedicalDevices/Safety/AlertsandNotices/ucm535843.html

[22] http://www.tomsguide.com/us/bluetooth-lock-hacks-defcon2016,news-23129.html

[23] http://fortune.com/2017/01/29/hackers-hijack-hotels-smart-locks

[24] G. Hernandez, O. Arias, D. Buentello, Y. Jin, "Smart Nest Thermostat: A Smart Spy in Your Home", Black Hat USA 2014, https://www.blackhat. com/docs/us-14/materials/us-14-Jin-Smart-Nest-Thermostat-A-Smart-Spy-In-Your-Home-WP.pdf

[25] http://ns.umich.edu/new/multimedia/videos/23748-hacking-into-homes-smart-home-security-flaws-found-in-popular-system

[26] http://www.gartner.com/newsroom/id/3165317

[27] http://spectrum.ieee.org/tech-talk/telecom/internet/popular-internet-of-things-forecast-of-50-billion-devices-by-2020-is-outdated

[28] https://www.shodan.io

[29] http://www.cisco.com/c/en/us/about/security-center/secure-iot-proposed-framework.html

[30] https://www.iiconsortium.org/pdf/IIC_PUB_G4_V1.00_PB.pdf

[31] https://www.dhs.gov/sites/default/files/publications/Strategic_Principles _for_Securing_the_Internet _of_Things-2016-1115-FINAL....pdf

6

Security in the Industrial
Internet of Things

**Aske Hornbæk Knudsen, Jens Myrup Pedersen, Mikki Alexander,
Mousing Sørensen and Theis Dahl Villumsen**

6.1 Introduction

In the previous chapter, security in relation to Internet of Things was
discussed along with different application areas. This chapter looks specifically into industrial applications of Internet of Things (also known as the
Industrial Internet of Things) through a case study of a Smart Production
platform. The Industrial Internet of Things is part of a current trend with more
and more industrial devices and systems becoming connected to the Internet,
including robots and "smart production" systems. An example of this is the
German initiative called Industrie 4.0 [1] backed by the German government,
followed by other especially European countries with similar initiatives such
as the Danish MADE [2] and the French Industry of the Future [3]. However,
too often the security of these devices and systems is either completely ignored
or not taken sufficiently serious. This assumption is supported by reports from
both government offices and security companies, but as attacks and events
where companies are compromised are considered confidential and sensitive
there are probably large shadow numbers.

Industrial systems are among the most critical systems in society, especially compared to private users. First of all because production facilities are
critical for a society to operate, but also because the economic consequences
of many different kinds of attacks can be severe: Whether it is about stealing
information, leaving systems inoperable, or even physically damaging systems
(for example, imagine industrial robots moving in ways so they destroy
themselves and each other).

In this chapter, we describe how we analyzed an automated production
line for security flaws, and based on this analysis became able to perform

a number of undesired activities. Among the flaws discovered were an SSH (Secure Shell) service with support of RC4 (Rivest Cipher 4) encryption and the use of default passwords to various services. Also, the central computer (control system) is running Windows 7 pro and can be accessed through Remote Desktop. For each of the mentioned flaws/exploits we discuss how to improve the security related to it, and so the chapter is concluded with a list of recommendations for designing, implementing, and configuring such systems. The production line used was acquired by Aalborg University in 2016 as part of a research project on Smart Production. In the current setup, it is assembled and configured "out-of-the-box", and no special care has been taken to properly secure it.

During the chapter, we demonstrate the consequences of setting up a system without taking security threats seriously. In fact, we were able to accomplish the three goals set up in order prioritized after how critical they are: (1) add/change/delete orders, (2) obtain performance information data as well as information on production costs without authentication, and (3) to cause severe damage to the system by deleting large number of vital files. In addition to the case study the chapter is concluded by a set of guidelines/ recommendations based on the observations of the study.

While such systems should not be used in production environments before being properly configured and secured, we believe the approach is not completely unrealistic. This is partly based on the observation of security in Internet of Things, but also backed by a recent study by SANS on attacks against SCADA (Supervisory Control and Data Acquisition) systems. In this study, 17% of the 268 participants had no systems to detect vulnerabilities, and approximately 30% of the participants consider attacks coming from within the internal network as one of their most vulnerable attack vectors [4].

The chapter is based on work by students at Aalborg University [5].

6.2 Background

This section will describe the system which was tested.

The authors were deliberately not given any documentation about the system, only a brief introduction on how to operate it. All information about the system in this chapter is based on research, assumptions and educated guesswork by the authors.

The system is capable of creating one or more products at the same time, using the same machines in different ways, thereby increasing productivity and efficiency. A 3D (three-dimensional) rendering of the system can be seen

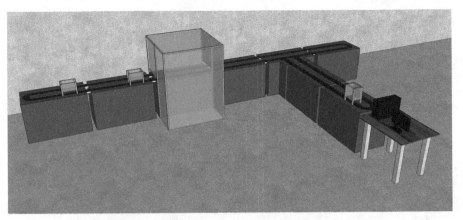

Figure 6.1 A graphical display of the production line.

in Figure 6.1. The system consists of units with belts, allowing production plates to move around and pass the different machines, so that a product can be produced. In this version, there are three machines mounted for production as well as a camera for quality control. All units are capable of doing an action, e.g. the belt, the machines, etc., and each unit is controlled by a PLC (Programmable Logic Controller), which communicates with the control system through Ethernet. The control system is a workstation running Windows 7 Pro SP1. A custom program, written in C# is running on the control system, with the purpose of controlling the production line. The rest of the chapter will focus primarily on the PLCs controlling the belts and network communication between the units and the control system, but the recommendations will be general as many of the security holes were found on all of the PLCs.

The control system has two Ethernet connections, one is directly connected to the Internet with no firewall, except for the build-in Windows firewall. In this setup, it is connected to the university's network and is therefore behind the university's firewall. The second Ethernet port is connected to the production line, and the two connections are internally bridged. It is not known at this point whether this was done on purpose or by mistake and never reversed, but in any case, it allows the devices/PLCs on the production line to be accessed from the outside network, which is a serious security flaw.

Each production unit has an 8 port Ethernet switch and the units are wired together, so that all units can access the control system. The production units are also equipped with another microprocessor and connected via Ethernet.

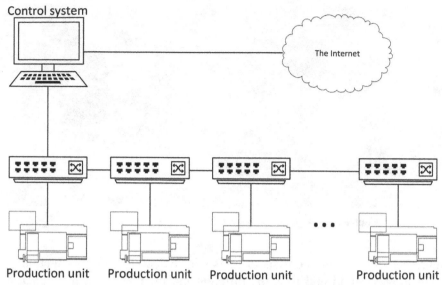

Figure 6.2   The existing network setup. It should be noted that the bridged connection at the control system PC, which bridges the internal network to the Internet, is not shown.

These additional microprocessors handle a GUI (Graphical User Interface) via a touch-screen. The versions vary a lot, from custom Linux on kernel 2.6 to Ubuntu 14.04 on kernel 4.2 and Windows CE 6.0 to Windows Embedded 7, of which only Windows Embedded 7 is still supported. The MES (Manufacturing Execution System) software running on the control system has an overview of the units, showing what state they are in, e.g. ready, working, or error. This is achieved by the units and the control system broadcasting some information. At this point, it is known that they are sending the last two octets of their IP address. When a production plate passes a production unit, the unit makes a request directly to the controller, asking for information on what to do with that exact production plate. If the answer does not concern this unit, it will let the plate pass and wait for the next plate. The current network structure is depicted in Figure 6.2.

6.3 Introducing Penetration Testing

Only knowing the background information summarized above, the first part of the study was to learn more about the system, both through research studies and by practically testing out the security of the network and different devices on it.

Penetration testing is used here as a method to discover weaknesses in IT systems and thereby be able to secure the systems against exploits or attacks. The process includes searching for vulnerabilities and often a number of proof of concept attacks are done as well [6]. Examples of such attacks are:

- Distributed Denial of Service: A DDoS attack is an attack where the victim typically gets flooded with either a large amount of random data or a large amount of connections. Both are done to saturate the connection to the server, making it almost impossible to be used. This is done to deny access to e.g. a website or other web-based services [7].
- Man-in-the-Middle: Man-in-the-Middle attacks are based on intercepting traffic. If an attacker can ARP-spoof (Address Resolution Protocol) both the router and the client in a network, it is possible to intercept the traffic between the client and the Internet [8].

This chapter will not describe either of these two methods further. While they could have easily been used against the system it turned out that less offensive methods were sufficient to reach the objectives set up for the study.

6.4 Methods

This section will discuss the methods and tools used to perform a penetration test of the system. The methods are divided into four steps: Describe the approach, Research, Scan and Find. These are explained in the following

- Describe the approach: The first step is to describe how the analysis is carried out, and which steps are taken afterwards in order to perform the penetration test of the production line.
- Research: In this step, as much information about the product as possible is gathered. It can be from physical inspection, speaking to users, reading through manuals, and finding relevant online information. As this is a pre-production model, there was no information about this model yet, and the manual was not available from the maker. Therefore, nothing was learned here initially. A brief introduction to the system was given by the owners, but it did not reveal much. After the scanning was performed, information about different parts of the system made it possible to re-visit the research phase and find relevant information about these parts.
- Scan: In this step a systematic port scan was performed on the entire /15 subnet. This was done in order to get an overview of the machines in the production line. Afterwards Wireshark was used to sniff traffic while an order of 10 products was being created by the production line, in order

to monitor what data could be observed during normal operation of the system.

- Find: In this step the findings from the previous steps are presented. This is done in the text below.

The scan revealed more open ports and more information than any penetration tester could ever hope for. Port 23 was open on 14 IP-addresses out of 14 in the entire system. In all 14 cases the machine was hosting a telnet service. This was also the case for port 21. In some of the 14 cases the machine was hosting a FTP (File Transfer Protocol (RFC354)) service. All machines have an open telnet and some an open FTP service. The telnet connection is of course unencrypted and therefore the usernames and passwords can be revealed through network sniffing, or possibly through brute-force attacks.

The network scan and traffic capture showed plenty of information about make, model and function of the different devices. It also, almost, mapped out the network, as all the devices in the production line broadcasted their function and other information at a steady interval. Before proceeding further, the sniffed traffic and scan results were reviewed and additional research was done based on the findings. Here a manual for an older model of the production line, containing the same PLC, was found, where the default password for the root user was given. It is not known whether the buyer is supposed to change this after the installation or if it is just left like this for easier maintenance. In any case it was not changed, and the old manual said nothing about changing it.

6.5 Tools

To find open ports and other vulnerabilities, multiple tools have been used. The tools represent standard tools for penetration testing, and what they find can be relatively easily found also by either security professionals or hackers with malicious intents. First of all, Nessus [9] was used to scan the most popular ports and find any vulnerabilities. Nmap [10] was later used to do a complete scan of all TCP and UDP ports.

When an open port was discovered it was tested for:

- Web interface.
- How the port reacted to establishing a raw connection with TCP or UDP.
- How the port reacted to establishing a Telnet/SSH connection.

The raw connection was especially useful to test proprietary protocols and test for false positives. To do these tests Putty [11] was used.

To analyse traffic to and from the devices, Wireshark [12] was used. Wireshark logs all the traffic on a given network interface and enables e.g. sniffing of login credentials to telnet. This requires the attacker's computer to be either bridging the user and device networks, be connected to a HUB between the user and device, or perform a Man-in-the-Middle attack like ARP-spoofing.

6.6 Findings

This section will go over the found security breaches and what further information this gave access to.

During the scanning an open telnet connection was discovered. This allowed access to the entire system as root. It was also possible to connect to the SSH service, with no brute-force protection. The SSH service allows the user to connect with an RC4-40 bit encrypted connection. RC4 is also vulnerable to a "Bar Mitzvah Attack" [13]. This encryption scheme is known to be weak due to its short length, and with today's computational power it is considered a trivial and fast task to crack. To our dismay the default passwords in all of the systems were not changed, and as a result of that it was possible to gain access to the entire system in under a minute with physical access to the network. The default passwords were found on Google after a quick search.

With root access to the units it was possible to get a look under the hood and learn more about them. It was discovered that the units are shipped with linux kernel 2.6 and Busybox. Busybox is used to handle most of the UNIX utilities. Both wget and GCC are installed so if needed it would be possible to download and compile other programs.

It was discovered that the access to the Internet is blocked by setting the default gateway to a non-existing IP-address on the network. This can easily be circumvented by setting up a device and assigning the IP of the gateway to that device and use it as the gateway. In the current configuration, this bypass is not needed however, because all of the units can communicate with the world through the control system because the connection between the two NICs is bridged (of course the standard gateway needed to be set).

Furthermore, during the scan it was discovered that most of the units were hosting some sort of web server on port 80, 8080 or some, rather odd, port, some had an FTP server and some allowed connections through SSH.

Using or even offering telnet has been strongly discouraged for years, due to the lack of encryption and the ease at which credentials can be sniffed, yet all hosts had a telnet server exposed. The SSH protocol is a suitable replacement for telnet, but attention must be paid to setting and maintaining secure settings. In the current case, configuration of SSH allowed RC4 40-bit encryption, which has been broken. The server accepted no stronger encryption than sha512, which is not considered secure. Furthermore, no mechanisms to avoid brute-force attacks were implemented.

6.7 Results

Due to the use of the same default login credentials on all production devices, with the user being the privileged root user, full access is obtained directly. Full access includes read and write access to the entire file system, and full access to connected hardware. From a technical perspective, this level of access to a device is typically the ultimate goal for a penetration test as it means that all security of the device is broken or bypassed (except physical security perhaps). With this access, software can be installed and configurations can be changed as per the attacker's desire.

Since the database on the control system is an Access database it would be easy to write malware in a language like C# to add, modify or delete orders. C# would be ideal for a windows system, because it only requires .NET. It is also possible to completely hide the application by running it as a service from the system user on boot and furthermore hide it in the task manger. The application can also communicate with a command and control server by hiding the traffic among other network traffic or attempt to hide it as legal network traffic. The application can be run in a silent installer with no sign of the program installing itself. As an example, the installer can be hidden in a simple PDF file containing a JPEG-2000 image and use the TALOS-2016-0193 [14] exploit to execute the program. This PDF file could be delivered by e.g. a spear phishing attack. This is just an example of how the system could be infected. Just recently the DirtyCOW (CVE-2016-5195) [15] bug was disclosed which could also be used to transfer and execute the code.

To cause severe damage to the system, the malware could be used to deploy a malicious, self-spreading network worm to the production units. This worm could connect via telnet, transfer itself and its current IP, execute on the new unit and then run the "rm -rf /" command on the current. Then the new unit waits for the previous unit to begin broadcasting again and the cycle repeats itself. This would completely stop the entire production and the worm could keep deleting hard drives until the default passwords are changed or the communication protocol is improved.

6.8 Recommendations

This section is divided into two parts. First, recommendations are given based on the system studied. While some of these are related to the specific setup and may not apply to all industrial IoT systems, most are still quite general, and can at least serve as inspiration for setting up other systems. Second, more general recommendations are given and elaborated on. These are in line with the recommendations given also in the previous chapter.

Insecure services like telnet should be completely removed and replaced with secure options like SSH, in order to enable encryption. SSH alone is not secure but it uses encryption and is capable of using very strong encryption. As FTP has the same insecure traits as telnet, FTP should also be disabled and replaced with a properly encrypted alternative.

The scan results reveal many open ports, corresponding to running services. As exposed services are potentially vulnerable, it is generally recommended to review the list of services accepting network connections, and disable any that is not needed. An alternative or complementary method, is to deploy a firewall on the device, to block any unnecessary incoming and outgoing traffic. An obvious benefit is that this is a simple mechanism, but is has an important cost on device resources.

Regarding the network, shipping the system with a firewall would also be highly recommended. This firewall should by default be restrictive, only allowing communication from the control system to the order-server. This would also ensure that the units are not able to access the Internet. Manually setting the gateway to an IP that does not exist should not be considered a secure solution, and in fact we are still unsure whether it was implemented for security or just pure misconfiguration.

The current discovery mechanism cannot be considered secure. The units broadcast important information about themselves, and the others just remember that unit until an override comes along. This is neither secure nor efficient. A more secure and scalable way would be to have the units communicate directly with the control system what they do, what IP they currently have, information about other units they need to query for some information, what to do with the current production, etc. This would make it much harder for an outsider (or even for compromised units from the inside, if this should occur) to collect information or spread through the system. If a running key encryption is used to encrypt the communication this would also mitigate the risk that an attacker enters the network and tries to mimic one of the units. Making all communication direct and encrypted would greatly increase the security of the system, which also lays the ground for the next recommendation.

The current IP delegation is far from optimal, if expansion is taken into account. The authors recommend adding a DHCP server, capable of handling the /15 scope and with the possibility of adding static IPs so that the control system is predefined, but all others are random all across the scope. This would help with expandability and security, as the attacker would have to scan the network to find the units and not just know that a 172.x.x.1 is a belt and 172.x.x.50 is a camera. Of course, this will complicate how the units find and discovers each other, but if the recommendations regarding that matter also are implemented, this will be straightforward to resolve. This could also increase security by adding a static ARP-table, thereby hindering ARP-spoofing.

The authors recommend the following changes to networking infrastructure. First, add a firewall, here named *Firewall #2*, which only allows traffic to and from the control system and a server, here named *Order DB*, containing all the orders, nothing else. Second, add another firewall, here named *Firewall #1*, which only allows traffic to and from the *Order DB* and whatever service is used, where the users register their orders, e.g. a web-server. An important thing to mention is that both firewalls should have three interfaces; one for the incoming traffic, one for the protected site and one where the management interface can be accessed. This last interface, should only be in use when the firewalls are being set up, or if the web-server changes IP and a new rule has to be added. Otherwise it should simply remain unplugged, so that attackers will not be able to access it if they somehow breach a host on the network.

If the suggested changes in communication and discovery are implemented, there will not be a need to change how that part of the infrastructure is build, except adding the DHCP-server.

A graphical display of the proposed networking infrastructure can be seen in Figure 6.3.

For the last part of the section focus will be on more general recommendations. In total, 13 recommendations are given and elaborated on in the following. Many of the recommendations deal with securing the MES controller, which is not surprising as the MES controller is really the heart of the system. However, this focus on the MES controller should also be seen in the light of securing the network and the individual devices/PLCs whereby it is ensured that they cannot communicate directly between each other, and that they can only be accessed through the MES controller.

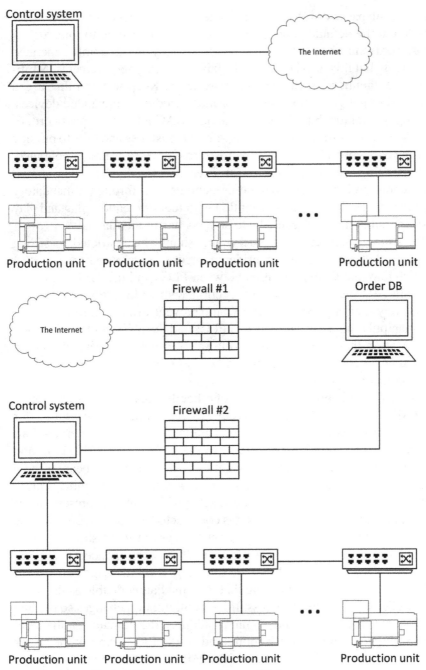

Figure 6.3 The proposed network infrastructure.

1. Use strong passwords for daily user interaction. A password should never be left at the default value, as it was shown to be trivial to find. Strong passwords must be hard to guess, so ideally they are randomly generated and long, but this is to be weighed against usability requirements. Multi-factor authentication can also be considered. Keep in mind that it can be hard to get good randomness for mass produced embedded devices. Using the unique, but non-secret, device MAC address as password, or deriving passwords from it, are well-known mistakes and fails to provide strong passwords. Reusing passwords across devices is also discouraged, as breaching one will mean that all are breached. Ideally the system should somehow prohibit use of default credentials, e.g. through a mandatory setup process. This appears to conflict with requirement for plug and play so a possible trade-off could be to address this in the manual.

2. Use strong encryption: All encryption should be restricted to strong ciphers that are currently considered secure, as opposed to e.g. RC4 which is weak. Communication between PLCs and the MES controller should be done using strong encryption with rolling keys rather than relatively shorter passwords. This would prevent an attacker from sniffing or manipulating messages between PLCs and the MES controller in case of a Man-in-the-Middle attack. It would also ensure that the attacker could not spoof the system pretending to be a PLC or even reverse engineer the protocol in the first place.

3. Optimize the method for discovery of other devices: In the current setup, the MES controller learns which devices are online in a way that is far from optimal, since all devices are sending with regular intervals four broadcast messages from port 1740 to the ports 1740, 1741, 1742 and 1743. It is unknown what the packet contains, since it appears to be a random string of text – the four packets appear to be very similar. This is not optimal, since it tells a potential attacker which units are present, and if the attacker knows how the packet is constructed he could spoof a device with his own machine and maybe get the system to crash if it cannot recognize the IP address as a machine. It is recommended to instead use a method where the PLCs automatically send messages directly to the MES controller – something that is immediately doable as the MES controller has a static IP address. If the configuration changes so that the MES controller obtains a dynamic IP address, the domains on the local network should be correctly configured so the PLCs can ask the router for the IP address of the domain of the MES controller. This would also allow for a much more flexible setup, where the whole network scope

can be utilized, and where more kind of tools/devices can be used since this identification would no longer be based on the IP address.

4. Disable services which are not needed and maintain security of those that are. In this particular setup, it was discovered that Telnet was open on all devices, most had SSH open with RC-4 encryption, and a few had FTP open. None of these services are secure, and should not have been enabled from the beginning. It is highly recommended to disable services which are not secure, as well as services which are not being used. If an unsecure service is needed, it is recommended to replace it with a more secure alternative or raise security through proper configuration. For example, Telnet should be disabled and replaced with SSH with a limitation on which encryption algorithms are allowed, so that old and insecure algorithms are not used. Another example would be to replace FTP with FTPS or SFTP, if there is a need for FTP even if SSH also supports file transfers.

5. Optimize the network structure: It is highly recommended that at least one firewall comes with the production line. If the MES controller is simply connected to the existing network in the building it is possible for any (compromised) machine to access the system. As depicted in Figure 6.3, we are suggesting to have two firewalls, where the Internet-connected machine that keeps track of orders can only communicate to the *Order DB*. Similarly, the second firewall should only allow communication between the MES controller and *Order DB,* something that would add complexity in how orders are communicated, but also provide a higher level of security against attacks from the outside. Both of the two firewalls should come with three network interface cards: One for incoming, one for outgoing and one for management of the firewall, where the latter is only connected if the firewall is actually being configured. This would ensure that an attacker, even if he could reach the firewall, will not be able to change its configurations without physical access. Moreover, it is recommended that the wireless network is either removed or secured through the use of WPA2-Enterprise and a RADIUS server – however, the wireless network should only be there if it has an actual functionality, not just for the smartness.

6. Use redundant hard drive on the MES controller.
7. Use external backup of the MES controller.
8. Limit the physical access to the MES controller.
9. Enable password on the BIOS for the MES controller.
10. Encrypt the hard disk of the MES controller.

11. Use different passwords for Windows and the MES program.
12. Encrypt the data base of the MES controller.

    Recommendations 6–12 are elaborated in the following: It is recommended to update the MES controller so it runs the newest updates for Windows 7 PRO, and also to install a decent security package. All settings should be carefully considered to ensure that it is not possible to for example make a Man-in-the-Middle attack on the Remote Desktop. It is also highly recommended to use a password for local machine access, so that even short term physical access does not make it possible just to use the computer. To avoid loss of data, and to reduce downtime, it is recommended to setup RAID 1 or 5 for the MES controller to avoid data loss if a disk for some reason fails – a high-end RAID controller would make it possible to exchange a disk even without turning off the MES controller. As a precaution against theft it is recommended to use Windows BitLocker for encrypting the disks in the machine, which would make the controller a bit slower, but also prevent an attacker from access the data in case the MES controller is stolen. As an additional precaution against loss of data or theft, it is recommended to have a backup server in a different physical location. A last suggestion would be to lock up the MES controller, as to avoid that someone can pass by and insert USB sticks. Some would also recommend to use glue to secure both used and non-used physical ports, but this is again a tradeoff between usability and security (it can make it unreasonably difficult to change e.g. mouse or keyboard).

13. Apply network monitoring and maintain logs: In the previous chapter, it was recommended to consider monitoring network traffic, and analyze the network traffic in order to identify malicious activities. This recommendation can be repeated here, and might even be more feasible in a system that only handles known machine-to-machine communication rather than a mix of traffic generated from machines and human beings: The more precise and narrow the benign traffic can be characterized, the easier it should be to set up rules and filters that catches abnormalities. That being said, it is of course also a matter of how many resources an attacker can devote to stealth operations. Keeping logs of events such as device access and authentication attempts are invaluable when trying to understand a breach after it has happened. To avoid entrusting an attacker with maintaining the logs on compromised machines, shipping logs off devices for retention can be considered.

6.9 Conclusion

In this chapter, we demonstrated how we were able to hack an automated production line, with almost no advance knowledge of the system. The automated production line was implemented out-of-the-box, whereas in a real production environment additional security measures would hopefully be taken. However, we do think it is problematic that we were able to fulfill all of our 3 objectives: (1) add/change/delete orders, (2) obtain performance information data as well as information on production costs without authentication, and (3) to cause severe damage to the system by deleting large number of vital files. Many of the vulnerabilities and flaws we found were due to misconfiguration (or lack of configuration), e.g. the use of default user names or passwords. A surprising use of poor encryption also contributed to making it so easy to hack.

It should be mentioned that the setup was not implemented in a real production facility, but as a first setup of a system used for Smart Production research at Aalborg University. We intentionally did this study before starting to configure and implement additional security features, which would be appropriate to do in a real environment.

We note that the system is riddled with security holes by default, and nothing stops the system from being deployed and used in this default, unsecure state. It appears that security received insufficient attention from the manufacturer. Consequently, the responsibility for securing the system is left to the installing/maintaining engineers and users, and depends on them realizing the need for security.

Finally, we should mention that much of our work was based on gaining physical access to the system, and thus limiting physical access is of course also an important countermeasure. However, the potential presence of inside attackers, the possibility that strangers can obtain access to the facilities, and finally the possibility that malware can be installed in the system and used for attacking from the inside implies that physical access alone is not sufficient to protect production systems from cyber-attacks.

To conclude the paper, our most important recommendations are summarized here:

- Do not use default usernames and passwords.
- Do not use unencrypted connections like Telnet.
- Only use strong encryption algorithms like SHA-256 or stronger.
- Do not reuse cryptography keys on IoT devices.

We believe that further studies are needed in this field. More case studies as the one presented here could contribute to an understanding of the current situation, but there is also a need for development of tools and methods for securing the systems, and for understanding the potential threats and their consequences.

As a last word of concern, one of the challenges for security in IoT in general is that devices designed for one usage over time become part of larger systems, and that might change their usage and role including the threat picture. For industrial systems, we see the same picture, and we see that even fairly old embedded devices which was never designed to be accessed from outside are being directly or indirectly connected to the Internet. It is strongly recommended that extreme care is taken when connecting "old" devices to new systems, and the previous recommendations on segmentation of systems are also here important to keep in mind.

References

[1] https://industrie4.0.gtai.de
[2] http://made.dk
[3] http://www.economie.gouv.fr/files/files/PDF/pk_industry-of-future.pdf
[4] https://www.sans.org/reading-room/whitepapers/analyst/breaches-rise-control systems-survey-34665
[5] A. H. Knudsen, M. A. M. Sørensen, T. D. Villumsen, "Hacking an automated productionline", student report, Aalborg University 2016.
[6] P. Engebretson, "The Basics of Hacking and Penetration Testing: Ethical Hacking and Penetration Testing Made Easy", Syngress Basics Series, Syngress; 2011.
[7] https://www.us-cert.gov/ncas/tips/ST04-015
[8] F. Callegati, W. Cerroni, M. Ramilli, "Man-in-the-Middle Attack to the HTTPS Protocol", IEEE Security Privacy 7(1), 2009.
[9] https://www.tenable.com/products/nessus-vulnerability-scanner
[10] http://insecure.org
[11] http://www.chiark.greenend.org.uk/~sgtatham/putty/download.html
[12] https://www.wireshark.org
[13] I. Mantin, "Bar Mitzvah Attack – Breaking SSL with a 13-year old RC4 Weakness", Black Hat Asia 2015. https://www.blackhat.com/docs/asia-15/materials/asia-15-Mantin-Bar-Mitzvah-Attack-Breaking-SSL-With-13-Year-Old-RC4-Weakness-wp.pdf.
[14] http://www.talosintelligence.com/reports/TALOS-2016-0193
[15] http://dirtycow.ninja

7

Modern & Resilient Cybersecurity
The Need for Principles, Collaboration, Innovation, Education & the Occasional Application of Power

Ole Kjeldsen

Microsoft, Kongens Lyngby, Denmark

7.1 Introduction

For many years, the IT industry have discussed how best to address the growing concern about data & IT security. In recent years the discussions have moved into both the board room[1] of most companies, into governments and also grown to be a concern of many citizens/individuals. Hacking has become a global enterprise. Companies, countries, and individuals are routinely targeted by nation states and shady corporations and not a week goes by without news of some terrible cybercrime, data breach or threat to the digital part of our modern lives. Mobile devices and digital services proliferate, and businesses, public services as well as most individuals have become ever more connected and dependent on the digital tools and this way of life. We have all become accustomed to the convenience and speed of these solutions, and while the benefits are naturally the driver for the digitization, it is obvious that it also opens the door, to more cyber risk.

The 'easy-to-use' digital services evolve not only for the law-abiding individual and business. Also the average cybercriminal, now has as easy access to cheap commercial and 'easy-to-use' cybercrime tools, as any individual has to lawful digital tools – a regular international cybercrime business, with professional marketing, support and even loyalty programs has emerged. The modern well-organized cybercrime syndicates are also agile enough to

---

[1]PwC Cybercrime Survey 2016.

Table 7.1   Data breach statistics

	2015	2016
Estimated number of customer records breached	160 mln	3 bln
Average number of days between breach & detection	229	140
Breaches using stolen user accounts	76%	76%
Average cost per breach	$3,79 mln	$15 mln
% of breaches involving human error or fraud	50	52

Sources: Ponemon Institute Releases; CSIS-McAfee Reports & Verizon Data Breach Reports.

simply change tactics and targets based on the current security landscape. For example, as operating systems becomes more secure, hackers shift back to credential compromise and third-party application exploits.

Financially over the past decade, this has resulted in an increase in the amount of money spent on cyber defense from less than $10 billion to roughly $70 billion[2]. This number should of course should be judged in the light of an estimated average cost per breach of $15 million[3] – something that in itself justifies the increased attention at all levels of an organization.

Finally and unfortunate for the overall level of trust in the digital life, the emergence of state-sponsored cyberattacks has since around the year 2010, been growing in numbers, in sophistication and in severity. Most recent most dramatic and broadly reported The Democratic National Convention hack[4] of 2016 and before that the Sony Pictures Cyberattack[5] which was reported to be behind the remarks on Cybersecurity in the '2015 U.S. Presidential State of the Union'[6].

But even if it is broadly accepted that the further digitization of our lives and businesses, includes this increase in overall risk, the apparent opportunities for further growth and value creation for both nation states, enterprises, smaller businesses and the individuals alike, make digitization almost a force of nature. Digitization is at the very least a force that should not be stopped, but instead it should be harnessed, controlled, utilized & optimized, guided

---

[2] Source: Symantec CTO Amit Mital at Fortune's Brainstorm Tech conference in 2015 in Aspen, Colorado, U.S.A.

[3] Source: Microsoft General Manager Microsoft Azure & Security Julia White.

[4] Washington Post 1st report on DNC email hack: https://aka.ms/dnchack1streport

[5] https://securityintelligence.com/news/cybercriminals-took-sony-pictures-entertainments-network-reports-claim/

[6] https://www.nytimes.com/2015/01/21/us/politics/obamas-state-of-the-union-2015-address.html?_r=0

by principles & policies – all to ensure that maximum value is derived at minimal cost for all. For all involved it must also without a doubt be a top priority to search for and establish a more modern & comprehensive set of solutions to the management and minimization of the ensuing cybersecurity risk. As the stakes rise, the new value generated in today's digital world can quickly be diminished or even eroded as a result of a single security breach – regardless who is to blame. Thus, cybersecurity and privacy in today's world is not just an IT issue – it's a leadership, governmental, business & citizen issue.

This chapter will evaluate both some of the larger trends driving the cybersecurity domain, as well as some of the emerging technologies and new paradigms that are likely to have the biggest influence on our digital lives and consequentially the cybersecurity environment. And finally, it is the lofty ambition of the analysis, to end up in a list of recommendations for how to improve and further develop a modern, trustworthy & resilient cybersecurity strategy to the benefit for all users of digital services.

7.2 Trends

Analyzing trends in cybersecurity is a crowded space, and there is no shortage of reports available. In the resource section of this chapter you will find a list of many of the most popular and comprehensive of the kind.

69% experienced cyberattack,
67% was exposed to ransomware or
similar ransom based attacks,
65% is more concerned about cyber
security today than 12 months ago
55% expect cybercriminals to be the
biggest threat (up from 40% in 2015)

PwC cybercrime survey 2016

In essence a list can of course never be complete and the following is subsequently only an attempt to synthesize the many available sources and highlight some of the overall trends. Most reports agree that:

- The number of incidents and severity of incidents are increasing, and accelerating
- The number and types of actors are increasing
  - Proliferation of 'easy-to-use' cybercrime tools introduce many new actors
  - Nation-states sponsor actors and openly develop offensive capabilities
- Human errors still contribute large portions of known data breaches
- The amount of digital data generated makes it both potentially more lucrative and more likely to 'escape in obscurity'
- High visibility in media, creates reluctance to sharing experiences

> *Of the nine basic patterns in security incidents, the top 4 patterns involve human error or misuse!*
>
> *Verizon Data Breach Report 2015*

- Perception is that full security is possible so if you are breached it is your own fault and you should be accountable regardless
- Crisis & PR management, has become a major part of incident response.
- Large differences in geo-distribution of malware, vulnerabilities, infection rates etc.

A 2016 cybersecurity trend[7] study shows that 41.8% of vulnerability disclosures are rated as highly severe and in combination with the statistics of the type of malware encounter rates as illustrated in Figure 7.1, it becomes quite troubling.

Professional environments (domain-based computers generally considered professional computers) typically implement defense-in-depth measures, such as enterprise firewalls, preventing a certain amount of malware from reaching users' computers and consequently tend to encounter malware at a lower rate than consumer computers.

Meanwhile, as shown in Figure 7.1, domain-based computers encountered exploits nearly as often as non-domain or consumer computers. Since exploits take advantage of disclosed vulnerabilities, it indicates this being the biggest and most severe issue for professional organizations and staying up-to-date with security updates one of the most important parts of their defense. Naturally securing ID and access credentials, data processing and general

---

[7] 2016 Trends in Cybersecurity: https://info.microsoft.com/SecurityIntelligenceReportData Insights_Registration.html

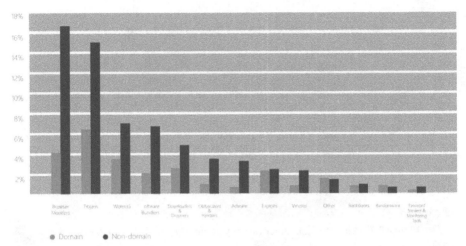

18%
16%
14%
12%
10%
8%
6%
4%
2%

● Domain   ● Non-domain

Figure 7.1 Malware and unwanted software encounter rates for domain-based and non-domain computers.

infrastructure by deploying modern cybersecurity tactics cannot be ignored either, but certainly good old-fashioned 'keeping you systems updated' remains a low-hanging, effectul and statistically significant cybersecurity measure.

> *More than 70% of attacks exploited known*
> *vulnerabilities with available patches. Some exploits*
> *dated back to 1999.*
>
> *Verizon Data Breach Report 2015*

What drives differences between the regions as seen below is out of scope for this analysis, but learnings from one region certainly can help other regions before attacks spread.

Ransomware is one of the most talked about cybercrime '*modus operandi*' presently, and is predicted to continue to evolve and escalate in the coming years. The notion that the criminal not only is unaware of the nature of data held ransom, but does not care, has taken many by surprise as they have unexpectedly become the target of the ransomware criminal. Most victims believed themselves to be uninteresting to the cyber-criminal, since they assessed that their data was of no to little value. While that might be true, it was of little importance since the ransomware attacks of course simply denied the data owner, access to the data rather than removing, destroying or copying it.

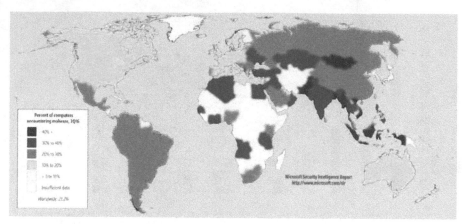

Figure 7.2   Malware encounter rates by country/region in 2Q2016.

Source: Microsoft Security Intelligence Report.

The reason for the increase in these types of attacks is really a sort of 'Perfect storm', where the attacks became very cheap and easy to operate, while many organizations simply misjudged the risk of being targeted. Thus, most attacks could have been avoided by well-known cyber-defense (attachment & URL scanning + user education) and recovery measures (Verified off-site Backup). Also for ransomware, rather large geo-distribution differences are apparent as shown in Figure 7.3.

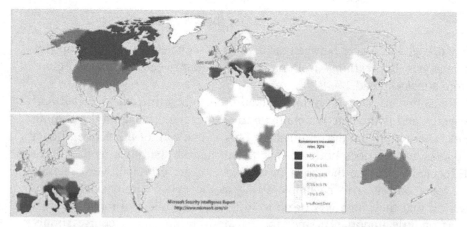

Figure 7.3   Encounter rates for ransomware families by country/region in 2Q2016.

7.2.1 Trends in Summary

1. Cybersecurity threats and incidents are growing in numbers and sophistication.
2. Most attacks (including ransomware) can still be avoided by following well-known advice and best practices.
3. Human error and/or simple fraudulent behavior (such as phishing) are behind most security breaches.
4. Countries and regions are very different in vulnerability profile and attack history/timing.

7.3 Protect, Detect & Respond

Before entering into any technical or procedural security control implementation under the classic Protect, Detect & Respond (including Recover) model, a couple of steps should be obvious to most security professionals.

You need to know your business, your data, your systems, procedures, legal requirements, liabilities and actors, and from that analysis get to an understanding of the context of the cybersecurity risks you face. At a much lover and much simpler level this goes for the individual as well and at a higher level for the nation-state.

This risk analysis is fundamental to aligning with the business and creating a proportional security strategy. In the conclusion of this analysis, you will see a list of suggested measures and initiatives – those should naturally only be contemplated *after* the completion of this basic exercise.

7.3.1 Protect 🛡

Once the preceding risk, data and overall environment analysis is complete and you have the overview needed, a natural first step is to go ahead and protect & monitor your infrastructure devices, network and physical environment, using the best practices readily available, e.g., in the Cyber Defense Operation Center Strategy Brief[8], and include Access Control, Multi-factor Authentications, Data Classification, Patching and Backup Procedures, Email and URL Scanning, Policy Enforcements, Just-enough Administrator Privileges Setup, Awareness Training Activities etc.

---

[8]https://aka.ms/mscdocstrategy

7.3.2 Detect 🔍

Next you need to consider that in a modern complex environment it is prudent to assume breach. This requires that you introduce behavioral analysis and tools to early identify deviations from your 'normal' situation, and any sign of compromise.

You should consider how widely your detection and analysis system is connected. Does it pick up not only on the known and 'close-to-home' attack types, but also allow for rapid update from a global large scale sharing data analysis and potentially the 'deep-learning' based detection capabilities of an AI based system. The latest in behavioral analytics similar to what the credit card industry over the years have developed and utilized in fraud detection – only applied to modern huge scale cloud computing capabilities. In the section on Emerging Technologies, such systems are discussed and you can read more about how Microsoft applies these technologies and procedures to their own infrastructure in the Cyber Defense Operation Center (CDOC) Strategy Brief[9].

And then naturally: when a notification that something abnormal in your architecture/network has been detected, it should trigger your response teams to engage.

7.3.3 Respond ☑

Finally modern organizations need to accommodate a growing need for responding quickly, limiting the impact of a breach, securing evidence and forensic data to assist law enforcement, limiting the potential liability for the organization and living up to any data controller responsibilities, such as the EU General Data Protection Regulation (GDPR[10]) Article 31. The need for forensics capable resources as well as supporting reports, log analysis tools etc. is obvious in this step. Again the CDOC[9] will offer inspiration.

7.4 Beyond Protect, Detect and Respond

7.4.1 Cyber-Offense

As mentioned in the Introduction, nation-states around the globe are evolving cyber-offensive means, and reports[11] indicate that some also are exercising

---

[9]https://aka.ms/mscdocstrategy

[10]http://eur-lex.europa.eu/legal-content/EN/TXT/?uri=uriserv%3AOJ.L.2016.119.01.0001.01.ENG&toc=OJ%3AL%3A2016%3A119%3ATOC

[11]https://www.wired.com/2015/09/cyberwar-global-guide-nation-state-digital-attacks/

this capability in cyberspace. While public proof of such endeavors are scarce, states such as the United Kingdom publicly state[12] their right to not only develop, but also bring these capabilities into action, should they chose to do so. They see this as a form of strategy of deterrence, and needed as a result of the UK being *"a hard target for all forms of aggression in cyberspace."*

Motivations for developing and activating such offensive capabilities have been analyzed extensively[13] – and it is not the focus for this analysis.

However it cannot be ignored that the mere fact that such capabilities are contemplated, let alone that they exist and are openly being deployed, presents the digital domain with a growing target on its pack. The risk of uncontrollable proliferation and escalation of the conflict between on one side value creation, economic growth & cost reduction, and on the other side cybercrime, erosion of trust in digital solutions and an unhealthy cyber arms-race, is very real.

"Force is all conquering, but it's victories are short lived."

Abraham Lincoln

A number of real risks ensues once you decide to build a cyberweapon or cyber offensive arsenal. Not only does history prove[14] that such a weapon will most likely proliferate and eventually end up in the cybercriminal toolbox and then only add to the problem, it is also a clear ethical question in need of contemplation and open democratic discussion.

As always with our state sanctioned handling of criminals, the issues are balance between proportionality, sheer retaliation or sense of justice, reformation and actual punishment of the criminal – and those are no different for any use of offensive cyber-tools. Add to that the risk of causing state to state conflict in cyberspace and the topic becomes even more worrying.

There is little doubt that the Internet, with its global connectivity, anonymity, potential to shelter yourself from consequences and the general lack of traceability, poses considerable challenges to those in the private and public sectors that are tasked with protecting it. The breadth of criminal activity, the number of actors and motives, and the lack of reliable attribution have all served to make crafting responses to attacks difficult. While there

[12]UK National Cyber Security Strategy 2016.
[13]https://www.fireeye.com/content/dam/fireeye-www/global/en/current-threats/pdfs/fireeye-wwc-report.pdf and http://www.cse.wustl.edu/~jain/cse571-14/ftp/cyber_espionage.pdf
[14]https://www.washingtonpost.com/news/the-switch/wp/2016/08/17/nsa-hacking-tools-were-leaked-online-heres-what-you-need-to-know/?utmterm=.559829592fda

are no easy answers, greater attribution and clearer rules for responding to both non-attributed and attributed attacks would enable the development and implementation of better strategies and tactics for responding to cyber threats.

But we should never feel compelled to impede the core principles of something that has proven such a source of tremendous value, by over showering it with suffocating defenses or soiling it with offensive retaliatory mechanisms.

We are at a point, where we need to accept the reality of cybercrime – that most cyber criminals will never be caught and some will operate with near impunity.

So instead of aiming to hurt/punish the criminal him/herself, it would seem we must do our best to deter and disrupt the business they run, and simply make it as hard as possible for them and their organization to succeed in their criminal endeavors. In essence make cyberspace a place where crime doesn't pay. Only then can we continue to enjoy the unprecedented force of value creation the Internet offers, and still protect our values, principles, property and ultimately our life and way of life.

7.4.2 Deterrence & Disruption

Over the past years, it has been proven time and again, that it is possible if not yet to DETER, then at least DISRUPT the growing business of cybercrime. This is done by a combination of modern cyber forensic tools and methods, well-known law-enforcement investigations, and finally close collaboration between public and private stakeholders resulting in take strong legal actions. No less than 14 botnets (see Figure 7.4) have been taken down as a result of this collaboration between private vendors and public sector authorities.

In effect the result is millions of 'bots' or zombie computers, being removed from the tool box of cybercriminals, which again results in a generally safer www and less cybercrime, to the benefit of all its users. This application of investigatory, judicial and law-enforcement power is possible even without the need for offensive cyber-capabilities, and does not carry with it any of the added risks discussed previously.

> *"A diplomat is generally cheaper than a solider!"*
>
> *unknown*

7.4.2.1 Resilience
It seems obvious that regardless of how we approach the modern world of digital services: protecting, detecting and even adding deterence and quick response to our strategy, we will not end up in a zero-risk cyber environment.

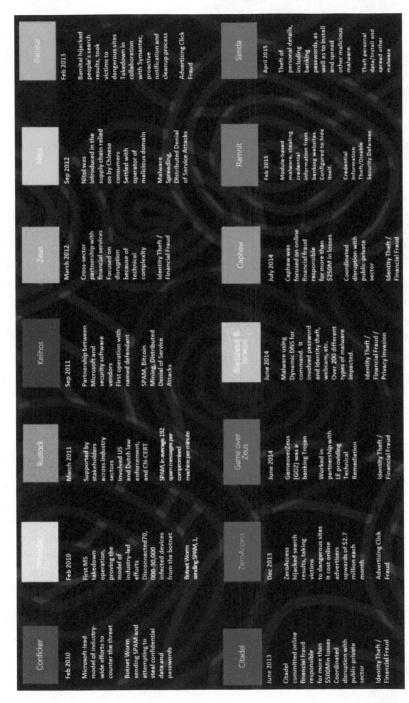

Figure 7.4 BotNet takedowns in collaboration between Microsoft and law Enforcement Agencies.

It is a natural consequence of the expansion of technologies into every aspect of our daily lives, that we will be exposed to cyber-risks, which cannot be fully mitigated. No interconnected system will ever be 100% secure, so vulnerabilities will exist and risks needs to be managed. If we then accept this risk and the fact that we cannot 100% secure all our interconnected devices, should we not match our preparations to this fact and be able to handle the consequences?

If so we need more than simply protect, detect & respond – we need to build the capacity to contain and recover. Or in short, we need resilience.

With resilience comes the agility (the adaptability and quickness) needed to survive a security incident, and the capabilities we build through our focus on modern detection, will then prove important for resilience. Compared to a classic defense-in-depth approach we have now secured the quickest possible reaction time and enabled many new procedural and IT architectural items that go into building the resilient cybersecurity setup – items as:

- *Containment*; the ability to isolate the threat
- *Forensics*; the ability to secure the evidence for further investigation, at best even without tipping of the cybercriminal, to allow further trace to be collected
- *Notification*; the ability to notify not only internal administrators, but also incident response teams and not least relevant authorities, in a timely manner and with relevant evidence.
- *Recovery*; the ability to continue business close to unaffected by the incident and potentially ongoing investigation.

But even this is not enough for true resilience.

At a foundational level, we need to embrace the theory of 'Adaptability' or 'Survival of the fittest' if you like and understand that unless we prepare ourselves and our organization for the unavoidable system failure, we will not be resilient and able to recover in any meaningful way from the shock & stress that will be inevitable from the incident bound to happen. Only if we design our thinking, our plans and systems to take advantage of the structural benefits of an interconnected architecture, where data and data processing can take many forms and routes, will we truly have built for resilience. The most illustrative analogy is really the classic web structure of the Internet compared to a more hierarchal and vulnerable setup.

After a catastrophic failure, the system and processes will need to be reinvented and find new ways to deliver the services that were disrupted and naturally learn from the disruption to find new ways to protect the new systems. In theory a self-healing system much like the Internet that can survive the loss

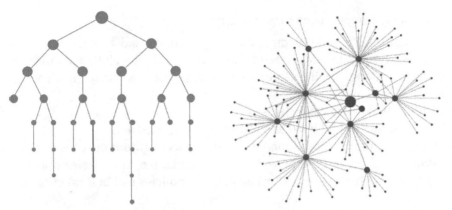

Figure 7.5 Resilience through system architecture – networked versus hierarchical flow.

of a number of nodes rerouting dataflows on the fly. We have then made use of the immense complexity inherent in the system and ensuing methods to deliver resilient services, and act as a complementary discipline to other modern cybersecurity efforts.

7.4.3 Importance of Culture to a Resilient Cybersecurity Strategy

Regardless if you believe in or decide to attempt to build the modern cyber-resilient and self-healing setup, there is no ignoring the cultural aspect of any cybersecurity strategy. As much as can be achieved with focus on technology and processes, much of it can be offset if no focus is given to the proper security culture around digital services/system, and the handling of data and devices. It is clear from the statistics (see figure below) that human errors happen at every level and need to be addressed from various angles to be effective.

TOP HUMAN ERROR SOURCES

42% End user failure to follow policies
and procedures,
42% General carelessness,
31% Failure to get up to speed on new
threats,
29% Lack of expertise with website
application and
26% IT staff failure to follow policies and
procedures

CompTIA Trends in Information Security report 2015

Constant investments need to be made in

- **Education**; both general digital literacy training and more complex role- & system based education as well as highly skilled focused education along the lines proposed in, e.g., the UK National Cyber Security Strategy[15].
- Identification of **best practices**;
- **Story telling**; around good and bad practices created and
- Delivery of **modern tools**; and pervasive **process guidance**, constantly **nudging** the user down the right path, to make the right decision and support the cybersecurity efforts and overall policies and best practices.

7.5 Global Security Intelligence Graph

In the globally interconnected world the amount of data known to be collected and analyzed on web user behavior & traffic is already extensive. It will by all accounts grow massively over the coming years. It is a of simple reaction to the growing amount of data on or associated with the web and the fact that more and more stakeholders realize the potential value to be found in analyzing these vast amounts of data, using the very same technology that enables the data creation in the first place.

Already today companies like Verizon, Symantec and Microsoft are building engines to feed the knowledge of the traffic in their own infrastructure and the web as such, into the products and services they offer. Companies are combining their data sources, tools and intelligence from multiple partners, which greatly improves the ability to detect threats earlier and respond faster. The insights gained from some of these vast infrastructures (see Figure 7.6), help build cyber threat intelligence and ultimately protect Internet users around the world.

7.5.1 The Use of Big Data

Generally Big Data refers to the use of a combination of data sets, structured and unstructured alike. The use of Big Data in the work to improve cyber-security is no different – while the true and complete list of sources remains unknown – for obvious reasons – these are some of the sources playing a role in the complex analysis:

---

[15]UK National Cyber Security Strategy 2016.

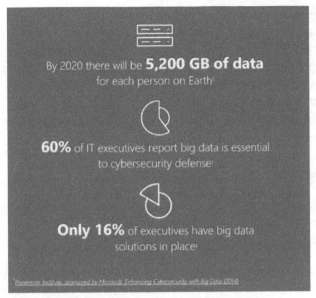

By 2020 there will be **5,200 GB of data**
for each person on Earth'

60% of IT executives report big data is essential
to cybersecurity defense'

Only 16% of executives have big data
solutions in place'

'Ponemon Institute, sponsored by Microsoft, Enhancing Cybersecurity with Big Data (2014)

Figure 7.6   The explosion of digital data.

- Signals from billions of devices,
- trillions of user authentications,
- scans of billions of emails and URLs for malicious content/intent,
- a constant flow of knowledge about current cyberattacks around the globe using known vulnerabilities, exploits and techniques,
- teams of white-hat hackers employed by cloud computing vendors, seeking and using latest tools from 'dark-web' sources to attempt penetration and not least,
- teams of professional 'hunters' also employed by vendors to defend and analyze attempts and reports from all available sources.

They all contribute to the detection of early signs of breaches and allow for rapid response before any real damage can occur and early protection when attacks spread to other regions of the globe.

The analysis of these huge datasets is more than most organizations can handle alone, but the arrival of Global Cloud Computing power such as AWS[16] and Azure[17] allows the creation of actionable intelligence and tools to help in the fight against the increasingly sophisticated cyberattacks.

---

[16]http://www.amazon.com/aws
[17]https://www.azure.com

To close the loop, the intelligence security graph created through these analyses is further deployed through post-breach security layers such as the Windows Defender ATP. This functionality is designed to leverage a combination of deep behavioral sensors coupled with the powerful global cloud security analytics, to offer modern and resilient detection of attacks – some known, some unknown, but detectable through the use of applied artificial intelligence. Find more on this in the section on Emerging Technologies below.

In any data driven scenarios, it is
important to remember these wise words:
"Not all that matters is measurable,
and not all that is measured, matters!"

7.6 Emerging Innovative Technologies

Many new and innovative technology trends add both complexity, new threats AND new capabilities in many dimensions of cybersecurity. Three of the most imminent and potent are Cloud Computing[18], Internet of Things[19] and Artificial Intelligence[20].

7.6.1 Cloud Computing

Some new threats as well as security benefits can be encountered when adopting solutions that are fully cloud based, or when connecting on-premises environments to cloud services.

Enterprise IT teams have established policies and procedures designed for enterprise infrastructure and applications, based on their decades of security experience dealing with on-premises threats. Many of these policies and procedures can be used effectively in public as well as hybrid cloud environments.

Also a typical global cloud computing vendor will offer a higher level of compliance with relevant international or industry specific security standards and a more comprehensive set of procedural controls, than it is practically or

---

[18] https://www.nist.gov/programs-projects/cloud-computing

[19] Defined as a system of interrelated computing devices, mechanical and digital machines, objects, animals or people that are provided with unique identifiers and the ability to transfer data over a network without requiring human-to-human or human-to-computer interaction.

[20] In this context and in the area of Cybersecurity, AI refers in to enabling computers to learn and take action without being explicitly programmed to recognize the exact pattern of an attack, incident or vulnerability.

economically feasible for any regular national or regional hosting/outsourcing partner, let alone a single customer maintaining a data center infrastructure of their own.

However, as organizations move workloads to cloud-based services it is important that security teams keep abreast of changes in their threat landscape brought on by the emergence of cloud computing.

New types of threats[21] can be related to characteristics of the public cloud only, or to issues introduced by connectivity between on-premises environments and the public cloud.

- Disclosing IP on public sites (such as GitHub)
- Pivot Back attacks (reusing data obtained from a public cloud attack, to breach on-premises environments)
- 'Man in the Cloud' attacks (criminal obtains access to the synchronization token and receives a copy of synchronized data without actually being present in the infrastructure)
- 'Side-channel' attacks (conducting local attacks, by deploying an attack virtual machine on the same physical server as the target)

It is important to note that while cloud computing does introduce a new attack surface and some new methods of attack for the cyber criminals, most often it also introduces a very professional protection as well as detection and respond mechanism/setup. A much more comprehensive and modern security setup than most on-premises environment are capable of obtaining and maintaining. For example very effective countermeasures are in place to mitigate the 'side-channel' attacks in some of the most popular modern cloud infrastructures deploying multi-tenant. Obfuscation techniques effectively eliminate the risk in those environments.

7.6.2 Internet of Things

The growth of the Internet itself and naturally the Internet of Things as a concept, increases the amount of connected devices at an aggressive rate. Many of these devices are unfortunately adding to the cybersecurity challenges, due to the lack of configurable security settings, insecure implementation of the internet-connected devices or simple absence of any security framework. Increasingly these devices are then hacked and used to carry out larger Distributed Denial-of-Service attacks – the most recent and widely reported

---

[21] Microsoft Security Intelligence Report Volume 21.

incident was the October 2016 Mirai IoT attack[22] where millions of 'Smart' or IoT devices, such as Smart TVs, routers, CCTV cameras etc. all were breached by the Mirai Malware and used in a botnet to attack and ultimately take down the DYN DNS service, and as a result interrupted the use of many high profile Internet services. Many owners of the breached devices were unknowing of their participation in the incident and devices may very well function fine, leaving them in theory open to be exploited in future attacks.

The increased cybersecurity risk from the IoT devices is impossible to overstate and with the introduction of so-called 'SMART METERS' for remote control of the home energy consumption, it may very well even accelerate in scale. Many countries have rolled these devices out (i.e. introduced by law in Denmark[23] in 2013, in Italy in 2005 and by 2013 in the Netherlands).

7.6.3 Artificial Intelligence

The use of AI in cybersecurity isn't science fiction. The ability of machines to rapidly analyze and respond to the unprecedented quantities of data is becoming indispensable as cyberattacks' frequency, scale and sophistication all continue to increase. Some[24] has criticized the industry for being overly optimistic on the promise of AI and Machine Learning in making sense of the volumes of often noisy data. The research being done today, however, shows that automated cybersecurity systems can do many things with only limited human oversight. Through neural networks, heuristics, data science, etc. systems are being designed to identify cyberattacks, to spot and remove malware, and to find ways to fix bugs faster than any human ever could. In some respects, this work is simply an extension of the principles that people have got used to in their mail-filters or firewalls. The sceptics focus on the potential for missing a signal (needle) somewhere in the vast amounts of data (the haystack), and while there are certainly no guarantees for neither a 100% success-rate nor notifications without false positives, it is obvious that it would be counterproductive NOT to apply the capabilities of Artificial Intelligence and keep improving the technology, simply because a manual only approach isn't feasible with the amount and growth rates of data.

---

[22] http://heavy.com/tech/2016/10/mirai-iot-botnet-internet-of-things-ddos-attacks-internet-outage-blackout-why-is-internet-down/
[23] https://www.retsinformation.dk/Forms/R0710.aspx?id=160434
[24] http://www.continuitycentral.com/index.php/news/technology/725-is-machine-learn...

That being said, it is only natural to consider the qualitative difference with the AI's "end game", i.e. having cybersecurity decisions taken by technology without human intermediation.

This novelty brings with it entirely new challenges. For example, what would legal frameworks around such cybersecurity look like? How would we regulate their creation and their use? What would we in fact regulate? There has already been some insightful writing and research[25] done on this, but for policy-makers the fundamental challenge of defining what an AI is and what it is not remains.

Without such fundamentals, even outcomes oriented approaches could fall short as there is no certainty about when they must be used.

In fact, AI technologies will be complex. Many government policymakers may struggle to understand them and how to best oversee their integration and evolution in government, society and key economic sectors. This is further complicated by the chance that the creation of AI might be a globally distributed effort, operating across jurisdictions with potentially distinct approaches to regulation. Smart cars, digital assistants, and algorithmic trading on financial markets are already pushing us towards AI, how could we improve the understanding of the technology, transparency about its decision making, integrity of its development and ethics[26], and the actual control of the technology in practical terms? But it is also critical to understand the role AI can and will play in cybersecurity and resilience. The technology is initially likely to be "white hat" enabling critical infrastructures to protect themselves and the essential services they provide to the economy, society and public safety in new and novel ways. AI may enable systems to anticipate and rapidly mitigate security incidents or advanced persistent threats. But, as we have seen in cybersecurity, we will likely see criminal organizations or nation states seek to exploit AI to evade cybersecurity defenses or even attack. This means that reaching consensus on cybersecurity norms[27] becomes more important and urgent. The work on cybersecurity norms will need more public and private sector cooperation globally.

On a more practical note, AI is not magical, but by simple overpowering of the amounts of data it can be put to good use in anomaly detection as

---

[25] Potential AI Regulatory Problems: http://philosophicaldisquisitions.blogspot.be/2015/07/is-effective-regulation-of-ai-possible.html and Regulating AI systems: https://papers.ssrn.com/sol3/papers2.cfm?abstract_id=2609777

[26] 6 Ethics Principles for AI to best assist Humans: https://aka.ms/aiethicsprinciples

[27] Microsoft Cyber Security Norms WhitePaper: https://blogs.microsoft.com/microsoftsecure/2014/12/03/proposed-cybersecurity-norms/

mentioned above. AI systems can build a baseline, defined by the history and then if a statistically significant deviation from that baseline is detected, generate an alert for a human to process.

Following the suggested ethical principles for AI, it can even offer transparency into how and why the result is considered an anomaly and offer the opportunity to tweak the algorithm if deemed necessary and prudent. A truly smart and AI based system will even take into consideration ambiguous activity, other supporting evidence either for or against notification, history of treatment of previous similar notifications and correlations with other detections, before actually deciding on notifying or not. In theory (and soon in practice as well) these systems will have a major advantage over regular signature systems by enabling detections to move past what is already known, and discover possible new exploits.

The amount of 'false positives' will in true AI based systems inevitably decrease over time as the self-learning kicks in and the amount of 'similar but previous unknown detections' start to appear. It must be expected to only strengthen the overall confidence in the detection system.

In conclusion, it is worth noting that despite the challenges posed by AI in cybersecurity, there are also interesting and positive implications for the balance between cybersecurity and cyber-resilience. If cybersecurity teams can rely on smart systems to play defense, their focus can turn to preparing to handle a successful attack's consequences.

Figure 7.7   The use of AI in cybersecurity.

The ability to reinvent processes, to adapt to "black swan" events and to respond to developments that violate the fundamental assumptions on which an AI is built, should remain distinctly human for some time to come.

7.7 Partnerships

The notion of an eco-system of cyber security partners is pivotal for the success in creating the trust needed in Internet driven solutions in the coming years. While the global cloud computing vendors such as Amazon and Microsoft certainly have a unique ability to synthesize threat data faster than any organization before, no one single organization can solve the world's security challenges. And no-one can expect to be able to – in a successful way – take on the task of constantly analyzing the vulnerabilities in the complex distributed network of solutions and the attack methods applied by a growing number of cybercriminal players. Players who in turn increasingly become more advanced and sophisticated in their approach. Regardless if it is feasible, the value from such a synthesis increases many-fold with multi-stakeholder involvement. In such an environment, there is an all but obvious need for Public-Private Partnerships, collaboration between commercial competitors, collaboration between customers and vendors, vendors and law enforcement agencies, broad deployment of self-sustaining defense mechanisms and not least a fruitful continuation of the current discussion at the governmental level regarding norms of behavior in cyberspace.

Stakeholders need to interoperate and share not only their data in a reactive way, but also let each other take advantage of what tools and methodology they each bring to the task. The days are gone where a vendor of software, network, hardware etc. could simply apply trustworthy methods to their code and/or device building and maintenance. Open source frameworks like the Microsoft Secure Development Lifecycle[28] are important and creates an obvious foundation for building trusted solutions, and in combination with an evolving broader operational security posture and sharing norm, it creates the foundation for a new dawn and hope for the cybersecurity future, even in the current 'under-attack' environment.

Alliances need to be formed among a variety of organizations as a foundation for the much needed sharing of information and partnering on solutions, that span across the relevant industries, including governments and

---

[28]https://aka.ms/sdl

law enforcement agencies. All to the benefit of everyone relying on these modern digital services.

A security threat that shows up in one company in one part of the world can benefit everyone on the system immediately. Intelligence feeds and is fed by all of the connected products creating a virtuous cycle.

7.8 Conclusion

Thank you for reading through to this point – assuming of course that you have not skipped the preceding many pages of analysis in this chapter – and who can blame you really, if what you are searching for is 'the silver bullet' of cybersecurity ☺

If you have read through the chapter, you will not be surprised to learn that this magic 'silver bullet' is in fact available . . . . . only it is really not one, but a cluster of bullets all connected and all relying on each other to combine into the best, most modern and resilient cybersecurity.

*In cybersecurity, as in many of the
more complex matters in life,
diversity is needed in all
dimensions!*

*Your protection, detection &
response, must neither be too male,
too pale, too seasoned nor can it be
too agile or too focused on too
narrow an attack vector or method!*

*Stay Alert – Stay Agile
Keep Moving Forward &
Deploy Comprehensive Methods*

All of the following elements and actions from stakeholders are needed, if we are to be able to respond to the challenge of cybersecurity.

It is indeed one of the most pressing challenges of our time, and if we want to continue harvesting the marvelous potential of digital transformation in both business, public, consumer and personal services, we need to approach it with vigilance, energy & investments.

Thus, the following are the suggested actions for all involved:

Table 7.2 Actions for the 'over-national entities' such as the UN or the EU

Suggested Action to Achieve Overall Modern & Resilient Cybersecurity
1. Engage in international dialogue designed to create international norms for cyber space behavior. Creating these norms will be as difficult as it sounds, but it is still both necessary and, ultimately, unavoidable. Absence of such an agreement, unilateral and potentially unprincipled actions will lead to consequences that will be unacceptable and regrettable

2. Agree on basic principles for
    a. Sharing data across borders while respecting local privacy laws
    b. Reforming of government surveillance

3. Drive for widest possible general harmonization and standardization.

Table 7.3 Actions for nation states

Suggested Action to Achieve Overall Modern & Resilient Cybersecurity
1. Subscribe to and share data and experiences through Public/Private Partnerships (PPP)
2. Consider the negative consequences of building cyber-offense capabilities
3. Balance national security with right to privacy – exceptions exist already where appropriate and additions should not be made casually or in haste, due to risk of erosion of overall trust
4. Adopt national laws that protect cyber space, build law enforcement capability and capacity, and support international efforts to fight cybercrime
5. Expand and adhere to the Mutual Legal Assistance Treaty (a.k.a. MLAT*)

6. Invest, invest, invest:
    • Broad digital literacy education, is important to limit the attack surface for social engineering, phishing attempts as well as maximize the use of good practices – experiences like the 'Stop. Think. Connect' campaign in the US have proven educational campaigns as important as technical security responses
    • CERT's & national competent authorities for coordinating cybersecurity incident response, are important for ensuring resiliency
    • Data protection agencies, need resources to achieve best possible balance between exercising authority and penalties with advice and practical guidance on data protection.

7. Nation states set the tone for the public discourse – make a public spectacle of victims of data breaches/hacking and risk all incentives to share and engage in open dialogue will evaporate.

*https://en.wikipedia.org/wiki/Mutual_legal_assistance_treaty

158 *Modern & Resilient Cybersecurity*

Table 7.4   Actions for state agencies at both national, regional and municipal level

Suggested Action to Achieve Overall Modern & Resilient Cybersecurity
1. Adopt modern technologies to realize productivity gains and better cybersecurity resilience
2. Accept and assume breach
3. Promote use of international open standards and best practices, to improve resiliency, easy comparison in commercial markets and open fair competition
4. Promote transparency in every way possible.

Table 7.5   Actions for businesses of all sizes

Suggested Action to Achieve Overall Modern & Resilient Cybersecurity
1. Adopt modern technologies to realize productivity gains, proper data hygiene and better cybersecurity resilience
2. Best practices such as network segmentation, visualization, least privilege principles, timely patching, log monitoring and multi-factor authentication, should be adopted broadly to minimize both the attack surface, the potential lateral movement of an attacker and the size of a forensics job, should a breach occur
3. Invest in your people – skilled analysts and data scientists are the foundation of defense
4. Regardless of any plan around protection & defense, accept and assume breach and subsequently monitor for abnormal account & credential activity
5. Prepare and build in-company data security culture and overall digital literacy, as users are the new security perimeter
6. Share experiences and subscribe to security networks in PPP.

Table 7.6   Actions for international SW, HW and services vendors

Suggested Actions to Achieve Overall Modern & Resilient Cybersecurity
1. Drive PPP and facilitate sharing also in-industry to allow Big Data integration
2. Promote and exercise full transparency where possible
3. Act truly global and allow intelligence to flow across the globe
4. Invest in emerging technologies innovation
5. Assist in development of strategies, policies and ethics codex for cybersecurity and emerging technologies
6. Continue to increase abstraction level of cybersecurity tools to promote increased ease-of-access and ease-of-use among both professionals and individual consumers.

Table 7.7 Actions for individuals

Suggested Action to Achieve Overall Modern & Resilient Cybersecurity
1. Stay alert and teach yourself – be digitally literate
2. Vote with your feet and your business – e.g. Only use services and products that promote and adhere to best security practices.
3. Remember and use the Top 10 rules* of digital life.

*https://www.sba.gov/managing-business/cybersecurity/top-ten-cybersecurity-tips and http://aka.ms/top10securitytips

Cybersecurity is for everyone!

Not just the IT departments or the national law enforcement agencies!

If we manage to execute on all or at least the majority of those actions, we have a chance to continue building trust in the digital world, which is not only important for global technology companies, but vital for ensuring that organizations and individuals everywhere can use technology with confidence and clear added value.

References & Recommendations for Further Reading

On Cybersecurity in General

- Top Trends in Cybersecurity 2016 eBook: https://info.microsoft.com/SecurityIntelligenceReportDataInsights_Registration.html
- Microsoft Security Intelligence Report – December 2016: https://aka.ms/cstrends2016
- Verizon 2016 Data Breach Investigation Report: http://www.verizonenterprise.com/verizon-insights-lab/dbir/2016/

On Use of Emerging Technologies

- Security Intelligence Graph: https://www.microsoft.com/en-us/security/intelligence
- Using modern Big Data tools and a progressive Security Model to combat modern cyber threats and adapt quickly: https://aka.ms/fightcyberthreatswithml
- Brad Smith, President, Corporate Legal Affairs: Cloud for Global Good – 2016: https://aka.ms/cfg

Security Advice

- Forbes Article: http://www.forbes.com/sites/tonybradley/2014/03/05/the-business-world-owes-a-lot-to-microsoft-trustworthy-computing/#5b5000293471
- Microsoft Operational Security Assurance Whitepaper: https://aka.ms/osa
- Microsoft Cyber Defense Operations Center Strategy Brief: https://aka.ms/mscdocstrategy
- The Human element of Cybersecurity: https://www.cybersecure.org/pages/resources/CompTIA-CyberSecure-Human-Error-White-paper.pdf
- The Register: "If this headline was a security warning, 90% of you would ignore it:" http://www.theregister.co.uk/2016/08/18/coding_pop_ups_hit_em_when_theyre_idling_university_boffins_say/

On Recovery & Resilience

- NIST 800-184 Guide for Cybersecurity Event Recovery (Learnings): https://aka.ms/mscdocstrategy
- Incident Response Reference Guide – February 2017: https://aka.ms/irrg

Partnerships, Future Policy & Strategic Considerations

- Developing a National Cyber Security Strategy Whitepaper: https://aka.ms/mscybersecuritystrategy
- "Should we retaliate in cyberspace", by Gene Burrus: https://blogs.microsoft.com/microsoftsecure/2017/01/12/should-we-retaliate-in-cyberspace/
- Partnerships with EUROPOL, FIS (Fidelity Information Services) a.o.: https://news.microsoft.com/2014/02/12/microsoft-enters-into-new-global-partnerships-in-fight-against-cybercrime/#sm.000013maeqhos 8fgxw9fwi2t0p7tj#j8EWiUAeVtaALhYL.97
- "Rethinking Cyber-threat" by Scott Charney: https://dumitrudumbrava.files.wordpress.com/2012/01/24-rethinking-cyber-threat.pdf
- EU Cloud Computing Strategy: http://europa.eu/rapid/press-release_IP-12-1025_en.htm
- Creating a Digital Geneva Convention: https://blogs.microsoft.com/on-the-issues/2017/02/14/need-digital-geneva-convention/
- "Cyberspace 2025: Today's Decisions, Tomorrow's Terrain." https://aka.ms/cyperspace2015

8

Building Secure Data Centers for Cloud Based Services – *A Case Study*

Lars Kierkegaard

Founder of InnoPaze

8.1 The Emergence of a New Industrial Era

The invention of the steam engine dates back to the middle of the 17th century and it marked the early beginning of the first wave of the an industrialized world. This was followed by the second wave in the end of the 18th century with mass production and the wide spread use of electricity. In particular, the invention of the light bulb by Thomas Edison in 1879 was a particularly ground breaking innovation of that time.

Computers and automation started to take the center stage in the 1960´ties as the third wave of the industrialization. The Apollo space missions, which took its beginning in 1961, used mission computers, which had the same computing power as a simple pocket calculator has today.

Now, the world is moving into a new industrial era, also called "industry 4.0". Industry 4.0 marks the fourth wave in industrial development where people and things become connected (Internet of Things) and a wide spread range of services are accessible on a global scale in the cloud. Today, almost all sectors of the society rely heavily on ICT. This is illustrated in the Figure 8.1.

Some consider industry 4.0 as a hype and buzz word without any substantial content. However, industry 4.0 is envisioned as an integral part of the infrastructure of a modern society.

Industry 4.0 defines business eco systems, which interconnect devices, equipment and people across traditional sectors such as transportation, healthcare, public service and the financial sector. We are already now starting to

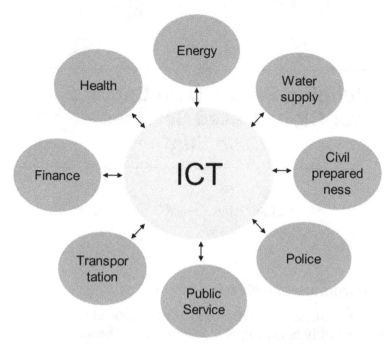

Figure 8.1   The modern society heavily relies on ICT.

see that more and more devices are connected and ITU-T is active in the standardization of an entire IoT framework [1]. Concepts such as the connected home are becoming a reality with different products entering the market such as temperature controlling heat thermostats and connected washing machines. Another example is connected cars, which we are starting to see on the roads today.

An important technology for making this a reality is cloud based computing, where experience gathering and experience based reaction is made possible everywhere. Various sensors detect changes in physical conditions, e.g., a rising temperature or low blood sugar levels of a human being. This triggers an alarm that is processed by a cloud application that decides to submit an alarm to the staff in a hospital operation center for further action. This is just one of many examples where Internet of Things and cloud computing hand in hand create new possibilities not seen before.

However, there are still a lot of challenges in the existing structures. In Denmark today, the police and civil preparedness use TETRA technology for voice- and narrowband data communication. The current ICT infrastructure is fragmented with a combination of both analogue-, digital proprietary- and digital IP based technologies. Examples include DTT (Digital Terrestrial TV) for TV broadcast, FM and DAB/DAB+ for analogue and digital radio, TETRA technology for professional radio for police and civil preparedness, DECT systems for health care personnel in hospitals and GSM-, UMTS- and LTE based cellular systems for mobile voice and data for commercial and private use.

8.2 Cloud Based Services and Data Centers

Cloud computing is defined as follows according to National Institute of Standards and Technology (NIST) in the USA:

> *"Cloud computing is a model for enabling ubiquitous, convenient, on-demand network access to a shared pool of configurable computing resources (e.g., networks, servers, storage, applications, and services) that can be rapidly provisioned and released with minimal management effort or service provider interaction."*

Cloud based services offer significant advantages over traditional client-server based solutions not only in terms of scalability in the number of users that the service supports but also in the ability to offer the services across all client devices, i.e. smart phones, tablets, laptops and desktop pc's. One of many examples of cloud based services includes Facebook, Dropbox, gaming platforms and more.

Cloud based services require that the frontend and backend software is hosted in data centers that support scalability in terms of CPU power, memory and data storage capacity. In addition, ultra-thin clients are supported across different client platforms such as smart phones, tablets, laptops and desktop PCs using a web browser.

Just recently, In January 2017, Facebook announced that they would implement a new data center in Fuenen, Denmark, which has the size of more than six football fiends. Other data centers are emerging in Denmark.

8.3 Types of Data Centers

Data centers can be dimensioned for different purposes and different service levels. Overall, data centers are defined and categorized into four different levels, or tiers, in accordance with the ANSI/TIA-942-A standard [3].

Tier 1 is the lowest classification, which offers a guaranteed service level of 99.671%. The capacity components of the data center, i.e. the uplink connection and the servers, which include the operating systems, memory and hard drives are non-redundant. This means that the capacity components are in fact single point of failure resulting in service downtime in case of software- or hardware failures.

The next classification level is tier 2 for which the capacity components must be redundant. This increases the guaranteed service level to 99.741%. In this configuration, the capacity components are redundant.

Tier 3 introduces dual-powered equipment, e.g. using UPS and diesel generators, and multiple uplinks to the data center. This, in turn, increases the guaranteed service availability with more than 0.2% to 99.982%.

Finally, we have the most robust configuration option in the ANSI/TIA-942-A standard, namely tier 4. In this configuration, all components in the data center are redundant. This includes HVAC systems, servers, cooling, storage and uplinks. In addition, the buildings are redundant e.g. in case of major disasters such as major fires, floodings, earthquakes and terrorist attacks.

8.4 Security Considerations

Different types of cloud services have different levels requirements pertaining to the categories of data centers. Typically, a tier 1 data center would be sufficient for hosting a gaming service, whereas a public safety & emergency service providing situational awareness for civil preparedness require a tier 3 or maybe even a tier 4 data center with disaster tolerance.

For all tiers it is important to secure the privacy of the customer's data as cloud services potentially include combining user data from different sources not only violating data protection rules as GDPR (General Data Protection Regulation) but also raising user concerns. This implies the implementation of additional security measures compared to traditional client-server systems.

8.5 Case: Teracom A/S

Teracom is an incumbent radio- and TV broadcaster with different facilities and assets in terms of 200 meter and 300 meter towers, nationwide WAN (Wide Area Network) infrastructure and buildings used for colocation of transmitter equipment, combiners and more [2]. The towers have associated buildings is depicted in the Figure 8.2. Many of the buildings date back to the 1960'ties or for some sites 1987 where TV 2 Danmark was introduced as a competing public service broadcaster to DR.

Figure 8.2 Typical 200-meter tower used for TV- and radio purposes.

Source: Teracom.

November 2010 marked the end of analogue terrestrial TV in Denmark. It was replaced with a nationwide DTT (Digital Terrestrial TV) system with a DVB-T (Digital Video Broadcast-Terrestrial) modulated signal. Now, it was possible to view more than 50 flow TV channels with a typical outdoor TV antenna.

The transition from analogue- to digital TV distribution left a lot of physical space empty in the buildings, which housed the analogue TV transmitters. The DTT system occupies only 1/40 of the analogue system, and the power consumption is much lower. At the same time, the eight main sites in Denmark are equipped with both UPS (Uninterruptable Power Supply) and diesel generators. This is the foundation for Teracom to enter into the hosting or data center business.

Teracom is one of the new players in the data center business in Denmark. The data center solution has been designed as a Tier 3 data center. I.e. the solution is based on dual-powered equipment, e.g. using UPS and diesel generators, and multiple uplinks to the data center. This configuration offers a guaranteed service availability of 99.982%.

Figure 8.3   Teracom A/S data center located in Hove, Denmark.

Source: Teracom.

The power supply to the data center is redundant and it includes an option for green power. Shared rack space can be provided for "smaller" cloud services. It is also possible to rent one reserved rack space, or an entire private suite.

Each rack is equipped with 2×16 Amps or 32 Amps in either 1-phase or 3-phases. The power supply to each rack is duplicated for each rack, and the power supply is secured with diesel and UPS in case of power failure from the external power supply.

An interesting option for mission-critical cloud services is an option to host the service is two data centers located in two different physical locations, e.g., east and west of Storebaelt. This offers disaster tolerance capability to the solution.

Continuous remote monitoring from Teracom's Network Operation Centre (NOC) of various censors in the data center is characteristic for a start-of-the-art data center today. Teracom's tier 3 data center supports continuous measurement of temperature, humidity and smoke detection.

Figure 8.4 The Network Operation Centre (NOC) is the focal point of any state-of-the-art data center.

In addition, the data center has a fire extinguisher system, which uses carbon dioxide gas that has a high rate of expansion. This allows the fire to be surpressed in less than 60 seconds.

A perimeter security zone is established through surveillance cameras, alarms, censors and even the possibility of physical guarding.

The existing AM, FM, DAB and DTT broadcast systems require only limited cyber security measures compared with data center solutions. For this reason, the IT infrastructure part of the data center is protected using firewalls and other mechanisms to protect the data center from cyber-attacks both with respect to denial of service attack, service authentication and not least service integrity.

8.6 Future Perspectives

Teracoms traditional TV- and radio broadcast business has been subject to disruptive changes in recent years. IP and the internet is the disruptive trigger, which has changed the traditional TV broadcast value chain completely and resulting in a more fragmented market with several new players entering the business.

The move from traditional broadcast to become a provider of data centers has increased the requirements in terms of cyber security by many factors and as a new area also necessitated focus on privacy concerns. For this reason, Teracom is in the process of obtaining an ISO 27001 certification in IT security. An ICT security policy has been defined and the preparedness plans have been expanded to also cover procedures in case of major cyber-attacks on the infrastructure and services. This in effect means a significant rise in security related investments.

References

[1] ITU-T Y.2060, Overview of the Internet of things. 06-2012, http://www.itu.int/ITU-T/recommendations/rec.aspx?rec=11559&lang=en
[2] Teracom A/S homepage: http://www.teracom.dk
[3] ANSI/TIA 942-A-2012, Telecommunications Infrastructure Standard for Data Centres, August 2012.

9

Pervasive Governance – Understand and Secure Your Transaction Data & Content

Kristoffer Rohde

Principal Sales Engineer, Opentext/EMC/Dell

9.1 Introduction

Effective in May 2018, the new EU General Data Protection Regulation (GDPR) will bring with it a series of obligations that requires organizations to rethink how they store, manage and report on the information they own or process as part of doing business. Even though May 2018 is quite some time away, many of the obligations in the GDPR will take time to prepare for and it can be a big scale undertaking.

Information is the lifeblood of any modern-day organization and as a result, information needs to be treated as a critical corporate asset and managed according to a well-defined information governance strategy. However, there are usually significant challenges when designing and implementing an information governance strategy for today's complex business environment:

- **Exponential growth in digital content**
  According to research from IDC, the digital universe is doubling in size every two years and will multiply 10-fold between 2013 and 2020 – from 4.4 trillion gigabytes to 44 trillion gigabytes. For perspective, today the average household creates enough data to fill 65 iPhones (32GB) per year. In 2020, this will grow to 318 iPhones. Systems need to manage this growth and simply "storing everything forever" is not cost-effective and it organizations must ultimately dispose of it according to consistent policies that both fulfill regulatory and legal obligations, and reduce information management costs.

169

- **A wide range of data & content types**
The face of unstructured information requiring governance includes, but is not limited to e-mail, file-systems, content repositories such as Microsoft SharePoint, EMC Documentum as well as user generated content residing on laptops/desktop PCs and mobile devices. Many types of information must be retained to fully satisfy compliance requirements: Transaction data and transaction logs, patient records, blueprints, vendor invoices and signature scans are just some of the items included in formal records for products and services.

  For full compliance, however, unstructured content is also required, ranging from sales presentations to emails, from customer proposals to written correspondence, and regulatory and legal requirements have expanded to include much of this content under formal management policies, and specific content types may have specific retention and disposition needs.

- **Legacy and active applications**
In many cases, particularly for recent transactions, records are on live production applications. In other cases, those records may be on legacy systems – the old sales system before customer relationship management was moved to the cloud, the old human resources system used by that company acquired in the late 1990s. And legacy information is everywhere: Human resource applications, e-commerce systems, customer sales and marketing databases. The amount of information stored is staggering, and generally must be retained for years. Retention policies are often mandated by governments, regulatory bodies, industry best practices and legal counsel.

  It is necessary to store content and data, which may span multiple obsolete or obsolescent applications, in order to enable the reconstruction of a specific transaction or event. Employee records may span HR systems, payroll records and emails. Financial transaction reconstruction may require not only transaction logs, but also historical pricing from stock exchanges, voicemails and bank receipts.

  Every IT department dutifully maintains that information, often by keeping old data available on new systems, sometimes by maintaining legacy servers, software and infrastructure for that purpose. Intelligent archival solutions can preserve and provide access to production and legacy information – combining data and content – at a lower cost, while improving access and simplifying regulatory compliance.

- **Increased pace of regulatory change – and tougher sanctions**
  Business and IT managers must respond to a wide range of rapidly changing rules and regulations that govern the management of information in order to address an increasingly complex legal and compliance landscape. Failing to address these obligations may result in significant sanctions putting the reputation or financial health of a business at risk. Compliance with retention requirements requires a lot of knowledge – and many IT resources. Internal controls staff require reports. External auditors require detailed records, including raw data and reconstruction of events. Litigators on a fishing expedition require everything.

This chapter focuses on how organizations can understand and secure their unstructured content and transaction data. This is the first step in applying a comprehensive approach to what is called pervasive governance.

9.2 The Challenges and Risks of Unmanaged Data & Content

Unmanaged file content lives throughout organizations on laptops, desktops, e-mail systems and – archives, on file shares, and in content repositories like Microsoft SharePoint and EMC Documentum. Transaction data, particularly for recent transactions, those reside on live production applications. In other cases, those transaction records or file content may reside on legacy systems, the old sales system before customer relationship management was moved to the cloud, the old human resources system used by that company acquired in the late 1990s.

Sometimes this information is project based and contains valuable intellectual property such as product plans, customer information, and more, and sometimes the information stored contains Personally Identifiable Information (PII) or Sensitive Personal Information (SPI). The result is thousands or millions of digital files scattered throughout an organization and transactions stored in legacy systems. Some of this data and content is important and some is not. Some of this data and content may carry risk to the organization if stored in the wrong location.

How do we identify which carry risk or are of strategic value? Would you have a person or team attempt to categorize it? Would you do the same, sorting through it, examining it, and categorizing it, confirming its value with others against a deadline, even though the stakes can be very high? With increasing regulatory pressures, possible security breaches and the threat of litigation, resultant fines, sanctions, and loss of organizational reputation, a lack of governance represent real threats to today's organizations.

It is critical for organizations to have good information governance in order to operate in today's complex business environment. If implemented without a well-thought-out strategy and design, governance practices have the potential for slowing down operations and inhibiting successful business results. However, a well-designed information governance strategy actually has the opposite effect by freeing users and organizations from the worry of policy compliance and allowing them to focus on delivering better products and services. Moreover, rapidly changing industry trends, such as cloud computing, are redefining the boundaries of IT infrastructures. With more data and applications moving outside of corporate data centers, information governance becomes the means by which businesses can move to the cloud with confidence that their content is being appropriately managed.

At the foundation of good information governance is knowing where all an organization's enterprise data and content resides and that its use is being carefully managed. As a result, it is a business imperative in today's environment that organizations ensure that their data and content is retained and disposed of according to corporate policies and/or industry regulations. This capability must both be broadly deployed and efficiently implemented.

Organizations can have a wide range of information retention and disposition requirements. Transaction data and content may be well organized, residing in an enterprise repository with readily available metadata for controlling retention. Or it may be unmanaged, spread around the organization in a variety of network locations and formats, live applications and legacy systems. Some content, such as contracts, financial records, or case management files, is mission-critical with highly regulated, long-term retention and access control requirements. Other content, such as Twitter feeds, blogs, or daily newsletters, may be transitory in nature with a very short shelf life and few privacy concerns.

Businesses have attempted to address this wide range of information management needs using a number of approaches:

9.2.1 The Fragmented Approach

In response to the growing volume and complexity of content that needs policy-driven retention management, many vendors have added an appearance of retention management to each of their products, forcing businesses to implement an overall solution by specifying content retention policies within each individual product. Each separate product then runs on its own, managing the retention of the content under its control.

Organizations that address retention management with this approach can encounter many problems, including:

- **Inadequate retention and disposition capabilities**: Many products do not have the necessary retention and disposition capabilities to satisfy compliance, certification, or other regulatory imperatives across all content, resulting in holes in a retention management strategy and a difficulty to enforce policies.
- **Inability to implement consistent retention policies across different products**: Each product has its own retention engine, with its own set of management interfaces. Attempting to coordinate retention activities across systems can result in retention errors, inconsistent results, and perhaps most worrisome, an inability to demonstrate compliance.
- **Inability to easily update retention policies**: Updating retention policies across multiple products can be time consuming, expensive, and difficult to accomplish in a rapidly changing regulatory environment.

This fragmented approach makes it difficult for organizations to scale their retention management capabilities to across the enterprise.

9.2.2 The Classic Records Management Approach

Classic records management systems provide extensive retention and disposition capabilities for content that is being managed to rigorously defined standards. Individuals, processes, or business systems can declare content as a formal record. The content is then managed by a records management system with a broad set of system-enforced capabilities, including the ability to:

- Specify and implement a file plan that details security, content permissions, retention duration, and disposition criteria
- Specify and implement a rich set of user and role-based record access controls
- Support and implement litigation "holds" on content that must be retained for specific legal or regulatory investigations independent of disposal status
- Monitor and manage audit trails detailing record possession, condition, transfer, and/or immutability

This type of rigorous records management approach is necessary for a subset of enterprise content whose management may need to conform to mission-critical, regulatory, or legal standards. Records management systems can enable companies to comply with wide-ranging recordkeeping requirements, reduce the cost of litigation and audits, and appropriately dispose of content once it has fulfilled regulatory and compliance obligations.

But this formal approach is not necessary for much of the other content in large, enterprise systems. This broader set of content needs appropriate retention and disposition controls in place, but may not need the additional controls required by formal records management. The time and effort required for both users and administrators to implement records management's broad, system-enforced capabilities can make it impractical for wide-scale deployment.

9.2.3 Keeping Legacy Systems Alive – Just In Case

To maintain access to the information residing on legacy systems, IT needs to consume considerable resources. Perhaps the legacy software application runs on obsolete hardware. The vendor supplying the legacy software may charge significant annual fees for licensing and/or patches. Staff must be trained and network infrastructure must be in place, as well as other physical resources such as racks, power, backup power, cooling and cables. Even though the information in a legacy system may be unchanging, there must be a data backup/restore system in place, as well as disaster recovery plans. After all, if you are required to keep systems running for compliance or other reasons, they must be protected as much as current line-of-business systems.

It's not rocket science: IT knows how to retain information for compliance. It's expensive, of course. Millions of Euro is spent each year to protect and maintain old hardware, software and infrastructure. Yet what is important to the enterprise, CIO and corporate counsel is the information, data and content. The end goal for such systems should be decommissioning, yet an end-of-life plan may seem to be at odds with compliance and governance policies. Data retention, however, is only one of the three goals of a regulatory compliance strategy: IT must also provide access and produce data.

Providing access can be a challenge, especially when access must be given to external entities such as auditors. Internal policies may prohibit giving access to production systems. What about when an audit or other query, such as e-discovery, requires access to many systems, many of which have their own user account schemes, or which require specialized access software?

Providing timely access to the retained data may be a bigger headache than the data retention itself.

Producing data on demand across those different systems is another headache. Need reports from an individual application? Easy. Need reports that correlate data across multiple applications? Hard. Pulling the data required to fully reconstruct a regulated transaction's many phases? Then you had better brace yourself for a tough ride, especially if that transaction contains both structured data (such as log files or accounting records) and unstructured data (like emails, phone calls, contracts and presentations).

In response to an audit request, e-discovery or government investigation, most organizations can provide reports that cover the basics of the transaction but are unable to provide third-party access or reproduce entire transactions, especially when those transactions took place more than a couple of years ago.

9.2.4 The Ideal Scenario

The approaches outlined above illustrate the fundamental challenge to retention management and comprehensive access to vital business information in an enterprise environment. How can robust retention management capabilities and access to business data across active and legacy applications be consistently implemented across the enterprise while simultaneously minimizing deployment time and cost, and the impact on user and administrator productivity?

Responding to this challenge in today's complex environment demands a modular approach to records and retention management. This approach avoids the pitfalls of a fragmented approach yet still provides the flexibility required to address a company's varying requirements across the entire enterprise. For an effective modular approach, a business should organize their records and retention efforts as well as efforts to ensure access to business data across active and legacy applications, around four main areas of capabilities: Enterprise Content Management, Core Retention Capability, Formal Records Management Capability and Archiving & Decommissioning.

9.2.4.1 Enterprise content management

First, consider whether to deploy an enterprise content management (ECM) platform suitable for wide-scale adoption to provide essential content management capabilities. This platform should feature an architecture capable of storing virtually any type and format of content. By utilizing a rich set of content services, security, and access controls, users and businesses can organize, control, and access their content through a range of client interfaces. By building upon these services, retention and records management capabilities become a seamless part of an ECM platform. This platform will form the basis by which an organization can design and deliver an information governance capability that is responsive to today's complex challenges.

9.2.4.2 Core retention capability

Second, strive to deliver a core retention management capability that can operate automatically, without user intervention, to consistently manage the retention and disposition of all enterprise content. This capability should be robust enough to satisfy the requirements of formal records management, yet be able to operate independently with a small deployment footprint in order to avoid the overhead of a full records management solution. Such core retention management capabilities should satisfy a broad set of regulatory, compliance, and investigative retention and disposition requirements for the bulk of a company's content.

9.2.4.3 Formal records management capability

Finally, design a records management capability that encompasses core retention management functionality as a "retention and disposition component," while also providing the complete set of additional services required for formal records management. The records management solution should have optional components in order to manage content not typically found in an enterprise content platform, such as physical records (paper or boxes containing physical records). This solution will be used for the specific content that must conform to mission-critical, regulatory, or legal standards.

9.2.4.4 Archiving & decommissioning – privacy by design

Every IT department dutifully maintains transaction data and content, often by keeping old data available on new systems, sometimes by maintaining legacy servers, software and infrastructure for that purpose. Intelligent archival solutions can preserve and provide access to production and legacy information – combining data and content – at a lower cost, while improving access and simplifying regulatory compliance.

Archiving that information into an intelligent enterprise archive, and then the decommissioning legacy systems can reduce both reduce cost and risk, while improving the usability and compliance of historical and valuable corporate information moving forward. Archiving not only eliminates the need for IT staff to maintain the old systems, but also means that information can be accessed in one place.

When information is migrated onto a consolidated archive, secure and compliant access becomes easy. No more terminal emulation or remembering the intricacies of an old version of shipping manifest solution. The information can now be made available directly to business and legal staff or even to regulators using a modern application with a modern user interface, in full compliance with industry and government regulations. Reports can easily be exported as raw data or PDF files, depending on what is required. The relationships between data sources are retained and can be explored through one interface, not dozens.

9.3 The Need for a Pervasive Governance Strategy

Creating a business environment that is cost-effective and empowers users, while also complying with regulatory and governance policies, can seem like an overwhelming task. Information, along with regulations, will only continue to grow. Doing nothing is not an ideal option. IT, compliance, and legal teams cannot just put their heads in the sand.

Today, many organizations have decided to "keep everything" forever. Doing so only increases the capital costs of buying more storage and the operational costs of having to invest more time and resources into tasks like backup and recovery. In addition, when the time comes to respond to an audit, investigation, or litigation, the task just becomes more challenging and expensive.

Implementing The Ideal Scenario as described above, for managing transaction data and unstructured content assets in an end-to-end content compliance platform across the enterprise, may not be a viable path for organizations in the short or medium term.

The GDPR will bring with it a series of obligations that requires organizations to rethink how they store, manage and report on the information they own or process as part of doing business, and they need to start planning what changes they need to make to address the new requirements.

Today, governance of unstructured content is not pervasive in most organizations. Oftentimes, organizations and their governance efforts are restricted to a particular content or application type that results in unconnected efforts and processes that end up as less than ideal. Unfortunately for organizations, governance is rarely embedded in the technology, processes, and business fabric like it needs to be.

In contrast, a pervasive governance strategy enables organizations to accomplish many things, including:

- Identify important information regardless of its location
- Classify and categorize it based on its business value and risk
- Secure access to content and protect it by putting restrictions on usage privileges, such as printing, copying, or sharing
- Make information accessible and shareable to end users, customers, partners, and others regardless of their location, even when it resides or is shared beyond the firewall and accessed via mobile devices

The best approach is to protect and secure information, while making it accessible and reusable based on policy and usage rights and permissions, regardless of its location. Today, locations often encompass on-premise applications and the cloud and information must be accessible via multiple devices which are increasingly a smart phone or tablet.

Policies, such as retention and usage rights, can help guide organizations that must adhere to regulations and internal mandates and help ensure adherence while still allowing end users to leverage unstructured content as a valuable asset. Setting policy is often the relatively easy part as many organizations have some basic policies in place. Connecting content with policy, given the scale and scattered landscape of the virtual organization, is the challenge most organizations grapple with. Being successful in this area requires doing intelligent work with fewer human resources with a premium on automation and transparency to end users in your organization.

Archiving your data and content into an enterprise archive, and then the decommissioning the legacy systems can reduce both reduce cost and risk, while improving the usability and compliance of historical and valuable corporate information moving forward.

9.4 Understanding Your Unstructured Content

When faced with the challenges of gaining control over their content, organizations are not always sure where to start. Most of us know that we have a lot of content and that it is located somewhere on the network, but how do we know:

- What file content we have?
- Who created it and when?
- What to keep?

- What to get rid of (delete)?
- What to secure?
- How to locate high-value records and high-risk content?

Answering these questions is the first step toward creating and executing actionable governance policies that dictate how long to keep information, when to get rid of it, who should have access, and more.

The challenge of course lies in the reality of today's content infrastructure. For most organizations, it is highly distributed, with remote workers and loosely managed environments like file-shares, e-mail systems and – archives and Microsoft SharePoint that hold large volumes of unstructured and unmanaged content.

9.4.1 Automated Intelligence

To address these challenges effectively and to be able to answer the above questions consistently and on a repeatable manner, there is an inherent need for automation and content intelligence. With the proper tooling in place, organizations obtain greater insight into the business value and risk of file-based content.

Multiple scenarios for discovering content environments are present. In some cases, metadata indexing might be chosen to produce the equivalent of a roadmap to uncover basic characteristics about content, such as its location, age, last update, and author. At the other end of the spectrum, full text indexing is similar to utilizing a Global Positioning System (GPS) to navigate your content infrastructure.

With full text indexing, organizations can uncover a greater level of file detail, which can be critical for identifying key business information that should be retained as a business record; or for responding to events like regulatory audits or internal investigations when identifying content such as payment card information (PCI), Social Security Numbers, taxpayer IDs, or intellectual property becomes critical.

These indexing choices are an important step toward understanding the business value of your unstructured content, illuminating what is valuable, identifying what may contain risk for the organization, and ensuring that this information is retained and protected rather than discarded.

9.4.2 Content Classification

Content Classification abilities should build on indexing efforts by categorizing an organization's content based on its value. Organizations can classify files by attributes (file type, owner/creator/accessed date/modified date) for foundational insight into content's business value. Content can also be classified based on its file content to identify critical business information that should be retained, sensitive information such as price lists and design documents, or to identify specific personal information such as PCI or PII (Personally Identifiable Information like Social Security numbers). Classification can also be based on combined metadata or based on user-created or administrator tags. These classifications and subsequent values can be applied to large volumes of similar groups of content. The results can serve as a catalyst for enabling a number of subsequent different policy actions and rules. This automated approach results in a quicker time to value and a higher degree of accuracy than the typical manual approach that most organizations are using today.

9.4.3 Actionable Intelligence through Reporting

Based on the indexed and classified content, organizations should be able to create summary reports that aggregates the index data as well as viewing detailed reports that includes metadata at the file level about each piece of content that fits the report criteria.

With either level of reporting, organizations can begin to make intelligent decisions concerning how to optimize their IT environments based on file information such as owner, resource consumption, age, and level of duplication. This information can be used to make decisions about what content to defensibly delete or move to tiers of less costly archive storage. In addition, from a governance perspective, reports are utilized to identify and measure levels of risk based on classifications that have been created.

Organizations may want to identify critical business information that should be retained as a business record or identify sensitive information such as payment card information or Personally Identifiable Information such as Social Security numbers contained in unsecured documents. This sensitive information may require additional security and protection corresponding with its deemed level of risk. By gaining visibility into the business value and risk of unstructured content, organizations can make appropriate and informed judgments to mitigate risk and maximize infrastructure investments to improve operational efficiency and reduce cost.

9.4.4 Automating Policy

Information from reporting should be used to develop policies and actions to manage risk in an automated or semi-automated manner. Content with strategic value or content associated with high-risk should be moved automatically to a secure location, where it can be effectively managed and leveraged across the organization, for example, as part of an ongoing business process or customer case file. When that content needs to be shared beyond the repository and network environment, policy that restricts usage rights, such as copying or printing, can be enforced to protect the organization throughout the content's lifecycle.

9.5 An Application Decommissioning Program

With an effective archiving & decommissioning program in place, organizations can reduce the cost of maintaining applications and static information, including structured and unstructured data, the contextual linkages between disparate data sources, and the software needed to provide access to that information. The archive preserves the unchanging information and its context, and provides easy access to it, rendering the original applications unnecessary and able to be decommissioned.

A typical organization may have hundreds of applications that store significant quantities of static information. It is not un-common that a typical mid-sized organization maintains more than 150 applications that are kept online, with all their data and context, and which are no longer in active use and, thus, could be archived and decommissioned. That usually means terabytes of legacy server software and databases and dozens of servers need to be powered, licensed, backed up, and maintained. The costs can be huge.

The first step should be to inventory the applications being maintained in your organization. The next step is to determine which of those applications are actively used but contain considerable static information that could be archived, and which applications are merely maintained in order to provide access to those old human resources records, product plans, log files from archaic applications, transaction logs from legacy sales platforms, and old customer-relationship databases.

Application decommissioning can reduce a company's dependence on application knowledge or on unsupported software products, operating systems, or hardware platforms. In addition, the company can use a well-planned strategy for application decommissioning to simplify its software and hardware infrastructure and centralize retention policies across decommissioned applications. Finally, application decommissioning can make it easier to mine data across applications, turn archives into information sources, and decommission relevant data subsets.

9.5.1 The Decommissioning Factory

Deploying an archive in a "factory model" is used to archive multiple applications, usually hundreds of applications, but in a strictly governed process. This model helps organizations to quickly and methodically identify and archive applications with a high Return on Investment (ROI). It can also help project team members hone their skills and make the archiving processes more efficient.

The factory model allows an established team to gain economies of scale. The team builds a body of knowledge that it can share with subject matter experts (SMEs) for the next archived application. Giving SMEs more time to gather information can speed up the archiving process and increase information accuracy. Archiving more applications also means greater cost savings and project ROI.

9.5.2 Developing a Roadmap

Experience shows that establishing a decommissioning program provides a powerful and effective program-governance strategy. Having a roadmap for building a decommissioning factory helps a business analyze the full impact of its strategy and is key to leveraging the solution fully.

9.5.3 Phase 1: Program Governance

Program governance may be provided through an existing project management function or through a newly created entity within IT. It includes operational guidelines for executing application-decommissioning programs and managing any overlap with other IT initiatives. Based on collaboration between IT managers and business representatives, program governance first seeks to define and develop KPIs and ROI metrics for evaluating the decommissioning program.

There are several staff roles that may be involved with a decommissioning project. Staff comes in and out of the project as needed and it is repeated for each application:

Role	Responsibility
Organization Executive Sponsor	Provides program vision and direction, resolves issues, provides direction and approval for schedule and budget changes
Organization Project Manager	Handles organization resource management, interface with Vendor project manager, and resolution of project management issues
Organization IT	Communicates architecture vision, provides system access, researches technical questions related to the source application, and helps resolve technical issues
Application Owner and Subject Matter Expert (SME)	Provides detailed information about the source application and defines high-impact business flows and high-impact reports
Vendor Program/ Project Manager	Handles Vendor resource management, budget and schedule tracking, and interface with the organization's executive sponsor and project manager and provides status reports on and facilitates the resolution of project management issues
Vendor Archive Architect	Manages the Archive installation, configuration, data loading, solution architecture, technical design, and quality oversight
Vendor Data Analyst	Collects data-retention requirements, data analysis, and inquiry and reporting requirements
Vendor ETL Consultant	Configures and executes ETL processes and validates source data extraction
Vendor Executive Sponsor	Provides direction, resolves issues, and approves schedule and budget changes

9.5.4 Phase 2: Application Decommissioning Factory Bootstrap

Establishing the application-decommissioning factory program includes the following tasks.

9.5.4.1 Train IT staff

The Vendor provides sample templates and a methodology tool to help train an organization's IT staff in application assessment. The vendor should also provide samples of completed application dossiers that show the information that is collected during application assessment and retirement. Staff training covers additional templates that facilitate program governance as well as the creation of decommissioning solution architectures, project testing plans, and deployment architectures.

9.5.4.2 Coordinate with other business services

Coordinating a decommissioning service model with other enterprise initiatives can help maintain a unified model for IT services. This typically involves sharing information in areas such as program governance, ROI models, KPIs, and project execution methods. By viewing information in a format that is consistent with other service offerings, business stakeholders can see a unified set of services and engage consistently with the staff members who provide these services.

9.5.4.3 Automate technology selection

Depending on the size and complexity of its application landscape, an organization may find it helpful to automate the classification of applications and the execution of decommissioning. In this phase of the rollout strategy, technology assessments and selections give IT the appropriate tooling to perform application assessment and decommissioning at scale.

9.5.4.4 Use proof of concept to reduce risk

As a company performs application assessments and identifies decommission candidates, it can discern any business risk by performing a limited number of proofs of concepts (POCs) in the early stages of the assessment program. The POCs can validate technology selection and synchronize the decommissioning methodology with other parts of the business. They can also speed the maturation of the assessment project and help the company identify technology and process gaps that should be resolved before full deployment.

9.5.5 Phase 3: Application Decommissioning Projects

Once the governance program for application assessment is operating, and an organization has established its roadmap for building the decommissioning factory, the organization can retire the original application and create functionality in the decommissioning archive. The organization will develop and analyze project KPIs to insure that the factory is generating the ROI that stakeholders expect and modify its decommissioning process as needed.

Application decommissioning projects typically involve four major steps:

- **Application assessment**, during which an organization identifies business applications that are candidates for retirement, based on specific criteria and projected ROI, and provides direction for the retirement project
- **Business and data analysis**, during which the organization sets functional requirements for the selected application(s)

- **Solution design and build**, during which the organization creates a customized, fully tested data extraction, transform and load solution based on their chosen archive solution and produces a solution-architecture document
- **Solution delivery**, during which the organization designs user-acceptance testing, user training, and knowledge transfer and then rolls out and deploys its solution

9.5.5.1 Business and data analysis

As organizations perform detailed business and data analysis for the applications they have chosen to decommission, the business analysts and technical leads for the projects should have detailed conversations with SMEs to determine which data to archive, how that data will be organized, and how it will be accessed.

The legacy infrastructure is then inventoried, including hardware, operating systems, and database platforms to determine access requirements for the ETL process. The user interface is analyzed to identify functionality that is still being used in the legacy application. Typically, for most decommissioning candidates, the user community limits interactions to read-only screens and reporting capabilities.

Finally, upstream and/or downstream application dependencies are studied to identify any data-flow issues that must be addressed through ETL. If a company expects to continue using upstream applications to flow data into the retired application infrastructure, it must have a clear picture of the data model impacts for the new archive. Similarly, the company should analyze integrations with downstream applications to understand how the data model works and ensure that the ETL process is complete.

The product of the business and data analysis phase is a set of functional requirements for the design and build phase.

9.5.5.2 Design and build

Using the functional requirements generated in the business-and-data-analysis phase, an organization then builds the decommissioning solution for the legacy application(s). The core infrastructure is assembled using a standards-based and vendor-neutral approach with the goal of providing a consistent application framework for consolidating possibly hundreds of legacy applications into a single technology stack. This consolidation allows the organization to outsource the legacy application access requirements and greatly lowers the overall TCO of the solution.

The design-and-build phase focuses on iterative development, testing, and deployment of the defined functionality. Plans for project management and change management (business readiness) are refined, and a detailed iteration plan and user-acceptance tests are developed for the current release. Refinements to the initial set of user stories and acceptance-test scenarios are created as needed, and the environment setup is completed.

During each iteration (typically two to four weeks in length), detailed development work is assigned, detailed unit tests and test scripts are created, and functionality is developed and tested according to these definitions. All iterations include internal quality assurance of the developed functionality, based on the detailed user-acceptance tests that were developed during the implementation prep iteration.

Business users should be directly involved in this testing, at least for every other iteration. Typically the final development iteration is reserved for solution hardening, documentation finalization, performance testing, and completion of any outstanding solution refinements.

In a typical project, a configurable user interface framework is delivered as part of the solution. As applications are assessed and decommissioned, the user interface is configured to add more search, display, and download configurations that will support accessing data from the source application. By consolidating multiple legacy applications into a single framework, an enterprise can minimize TCO and maximize its ROI as well as ensure consistent retention management across application data.

During this phase the ETL model is also completed and tested. Legacy data should be migrated into an industry-standard, future-proof XML format that is vendor independent, stable, and application neutral. Important aspects of the ETL include the timing, performance, and accuracy of the processing. Also important are the flexibility and scalability of the ETL design for adapting to the many unknowns and variances of older, poorly documented, or under-supported systems.

9.6 Conclusion – Solving the Challenges of Unmanaged Data & Content

Many organizations are engaged in high-volume records management exercises and programs. The reality is that these efforts can be quite challenging. Asking a group of dedicated records managers, or more likely end users, to classify and tag content as business records is both unlikely and not feasible in many cases and can create an undue burden on workers, negatively impacting worker productivity. Compliance groups also often assume that if there is a records program in place, everything that is a record has been identified and is under management. Rarely, if ever, is this true.

By implementing an automated intelligence framework that enables content classification, actionable intelligence and policy automation, the guesswork and end-user involvement are removed or minimized in categorizing content files that reside on the organization's file-shares, e-mail systems, laptops and desktop PCs and other systems. This process can happen transparently to end users, based on its classification, high risk or sensitive information can be moved to pre-determined secure locations and the appropriate retention and disposition policies are automatically applied.

The result is a reduction in time spent identifying important business records, greater accuracy and consistency in doing so, and the ability to make this a repeatable, systematic, and transparent process based on a schedule with predetermined outcomes. This process has the benefit of also being applicable to content that can defensibly be deleted upon its identification protecting the organization against unwarranted risk. Even in situations where some end-user involvement is required (such as in project-based work streams or verification of a certain content asset), the ability to automate much of the identification, classification, retention, and disposition requirements still works to improve consistency and minimize end-user burden.

Pervasive governance represents a fundamental strategy in helping organizations get the maximum leverage out of their information. The goal should be to enable organizations to do so while integrating governance and compliance into the technology, business processes, and business fabric of their organizations. For many organizations, the prospects of implementing a comprehensive approach to governance and compliance may seem daunting and even overwhelming.

This is the reason why this chapter has focused on providing insights and suggestions to help you understand the challenges and risks of unmanaged transaction data and content. The perfect first step is to assess where your current challenges lie, what content types are proliferating, and to determine their business value, cost, and potential risk. When implementing auto-classification capabilities that rely heavily on automation, your organization's productivity will increase.

Archiving your information into an intelligent enterprise archive, and then the decommissioning legacy systems can reduce both reduce cost and risk, while improving the usability and compliance of historical and valuable corporate information moving forward. Archiving not only eliminates the need for IT staff to maintain the old systems, but also means that information can be accessed in one place.

References

[1] The Digital Universe of Opportunities: Rich Data and the Increasing Value of the Internet of Things; *EMC Digital Universe with Research & Analysis by IDC, April 2014.*

[2] A 15-Minute Guide to Protecting and Controlling Content Wherever it Resides; *EMC Corporation, Enterprise Content Division, 2012.*

[3] A 15-Minute Guide to EMC File Intelligence: How to Understand and Secure Your Content; *EMC Corporation, Enterprise Content Division, 2012.*

[4] A 15-Minute Guide to Retention Management; *EMC Corporation, Enterprise Content Division, 2011.*

[5] EMC Perspective: Overcome Regulatory Data Retention Challenges with Compliance Archiving; *EMC Corporation, Enterprise Content Division, 2014.*

[6] EMC Perspective: Legacy Decommissioning – Good for the Budget, Good for Compliance; *EMC Corporation, Enterprise Content Division, 2014.*

[7] EMC Perspective: Application Decommissioning; *EMC Corporation, Enterprise Content Division, 2014.*

10

Challenges of Cyber Security and a Fundamental Way to Address Cyber Security

Fei Liu[1] and Marcus Wong[2]

[1]2012 Shield Laboratories, Huawei Technologies, Singapore
[2]Wireless Security Research and Standardization,
Huawei Technologies, USA

Abstract

There is almost never a day that the topic of cyber security is not on the front page of a leading newspaper or a prominent website. It seems to be a constant game of cat and mouse between the hacker community and the security community in terms of providing the protection for information, user data (corporate, financial, etc.) and user privacy to the point that the security community cannot provide hot patches and fixes fast enough. Adding to this is the undisputed evidential linking of many hacking incidents and capabilities to sponsorship from powerful and capable states that can provide almost unlimited attack avenues.

In the past, security controls and protocols for the Internet, communication, and systems are all practically built as an after thought and the need to have the system working and demonstrated trumps the need to make sure the system works in a secure manner. Functions over security, as many have seen and employed in the past simply does not work and it is the number one cause for many of the cyber security attacks that are on the front page today. Obvious, this model has not worked very well in the industry because we continue to see cyber security threat as the number one threat in many literatures and surveys. While recognizing this is a fundamental problem that we are facing for many years, why is it that this continues to be the number one issue, year after year? What is preventing the problems to be dealt with head-on?

There are countless of security professionals, organizations, and governmental agencies that are committed to deal with this and yet the problem persists. What is working and what is not?

10.1 Introduction

Many factors contribute to cyber security attacks in telecommunication systems and networks: lack of comprehensive threat and security analysis in initial design, poor implementation, rushed testing, careless deployment, etc. that result in weaknesses and holes to be exploited by attackers. All of these factors point to a break-and-fix philosophy that has not worked in the past. The lack of a complete and systematic approach to combating cyber security threats and attacks calls for a fundamental paradigm shift to a security assurance process.

Ever since the early days of electronic networking (e.g. telephone networks), various forms of cyber security attacks have been around. Back then, the attacks were simple, crude, but effective. The earliest such attack originated with the telephone systems. An attacker would climb up to the telephone poles, physically splice the wires together and call up the operator to place a call, as with all telephone calls that had to be placed by operators then. Since telephones were only available to the privileged few, there was no authentication of the caller as the operators would assume that the call had been originated from the person who had the telephone installed in his premise and then would connect the call as requested. As telephone technologies advanced, telephone system switches become automated and use of operators to place calls had been replaced by switches that can detect the number of electric pulses from the telephone set as the user placed a call via electric pulses or dials. Soon after, touch tone replaced rotary dial: instead of using the number of electrical pulses, special tones, or sounds that are generated on specific frequencies are used to represent the digit that is being dialled. The switch in the central office would interpret the sounds by the frequencies and place the call accordingly. Similarly, coin operated telephones in public telephone booths also use sound detecting technology, not only for the digit being dialled, but also for the coins being deposited as coins of different denomination made unique sounds by weight. When the sounds of the coin deposited is transmitted to the central office, the switch then authorized the call to be placed according to the amount of coins being deposited. Lacking cell phones in the old days, the public telephone made it easy for people to connect on the go, but it also made hackers life easier as well. Some of the more effective hacks on the public phone systems entailed making sound generating devices (e.g. Red Box) that mimicked the sounds of each coin being deposited in the coin-operated public phones. These attackers not only benefited from making free calls, but more importantly benefited from making calls that were untraceable while they went on to other illegal activities.

Fast track 30 years, the Internet boom and cellular technology has literally replaced most of the telephone system as we knew it then. This shift in technological advance along with security awareness, while thwarted the attack on some older and antiquated technologies such as the public telephone system has brought new cyber attacks that are much larger in scale and causes much worse damage than before. Stakes are much higher as the financial motivation of cyber attacks is much higher as well. Recently, the attackers in the 2016 Bangladesh Bank heist made off about $101 million (though some $38 million had been recovered) in one of the biggest cyber attacks in history. Other attacks are much more difficult to quantify financially as in the recent attack that saw hacked cameras and video recorders were used to launch several massive Internet attack that took down or slowed down a number of popular websites such as Amazon, Netflix®, etc. In terms of losses from potential revenue, the many websites that rely on revenue from traffic driven by the number of clicks on advertisement, number of products sold, other forms of redirect, and potential for future business opportunities are difficult to quantify. Costs related to recovery, reputation, potential product updates and/or recalls (due to vulnerabilities, incompliance, or faulty design, etc.) have a bigger ripple effect on the entire ecosystem that is far more difficult to quantify.

For every cyber security breach that is reported, there are many more that are not reported. After all, acknowledging such attack while not providing assurance such vulnerability has been fixed is yet another invitation for further attacks. Additionally, some victims took the approach that if the loss costs less than the fix; they treated the loss as part of the business. This was the case for numerous wireless operators operating first generation mobile networks where they viewed fraud resulting from cyber attacks as part of business.

Through out the years, the communication and networking industry, then the telephone network to the wireless to the Internet, all has been victimized as the cat-and-mouse game between the network operators and the cyber attackers continues. Products and networks would be built and deployed and services offered, at the same time, attackers have stepped up their games in seeking out vulnerabilities and launching newer, more sophisticated, and more powerful attacks. In turn, the product vendors, network operators, security experts, and even nation states came together to thwart off the attacks by plugging the vulnerability holes and add additional security measures. Driven by temptation of financial gains, notoriety recognitions, and other undisclosed reasons (e.g. vengeance), the attackers also put in additional effort, found new vulnerabilities to exploit and launched new attacks. Breaking-and-fixing as a norm between the good and the evil has resulted in a vicious cycle that calls out a new paradigm shift in this endless game. However, when some of the vulnerabilities and attacks are state-sponsored, it is a new ball game altogether: Snowden's revelation of massive abuse of power, privilege, trust by the Central Intelligence Agency of the United State, rogue states' hacking into other legitimate government and business' networks and systems, infiltration of computerized command and controls of utility companies in the U.S. and around the world further illustrates the need to re-evaluating the approach taken by the industry to combat these cyber security attacks. The break-and-fix paradigm that the entire cyber security ecosystem has focused on for many years simply did not work.

10.2 Security by Design

Network design, system design, and component design all are complex processes that take lots of time and efforts from planning to conceptualization to building to testing to deployment. The foremost importance is for the system to work as designed. Take the example of computer operating system (e.g. Microsoft Windows OS), it is an extremely complex software that has been developed by teams of hundreds even thousands of talented engineers, system designers, software developers, testers, system integrators as well as security experts. Chances are that the goals for the earlier versions of the software, as with many other products, are to ensure that the product works as intended, providing the input and output as the users are expecting. As with any other product development cycle, the "product" also goes through various and rigorous testing phases (e.g. functional testing, beta testing, boundary testing, fuzz testing, etc.) before being released to the general public. However, the test phases most likely were performed in friendly environments by friendly personnel according to the functions of the product as it is designed to perform in functional testing. In general, the software development process (as with other system development) goes through the phases of requirement, planning, designing, implementation, testing and document, deployment, and maintenance.

Fundamentally, this product development process (e.g. software development process) is a sound one from functional aspect as it is of foremost importance that the product works, can be deployed quickly, and start bringing revenue right away. However, as it has been proven time after time that not every user of the product is using the product as it is intended. Misuse, unintentionally or intentionally, can quickly send the product back to the drawing board. However, the developers are more likely to send the product to the maintenance phase of the product development cycle rather than the design cycle, saving time and money in an attempt to "fix" what is wrong and send the product back to the market as quickly as possible to keep the revenue flowing.

"Misuse", as it is put mildly, actually takes many forms. Although all forms of "misuse" have unintended consequences, however, the misuses that bring the most damages are in the form of malicious misuse. Malicious misuse has been motivated by a variety of reasons: financial gains, recognitions, etc.

Take the example of Microsoft Windows XP, designed to take advantages of growing use of graphical user interface in personal computers to make it more user-friendly after initial successes with earlier versions of other windows operating systems. However, as the product was built during the time when software development for personal computers was still in its infancy, many of the simple vulnerabilities and attack vectors that are known today, though already known to some software development community, were simply not taken into account in entirety. Microsoft Windows XP, before finally putting to rest in 2014 (i.e. stopping technical support), has been heavily criticized for the numerous security vulnerabilities from buffer overflow, malware, worms, etc. As the vulnerabilities were discovered, fixes or hot patches were put out through windows update to plug the holes. The automatic windows update feature of Windows XP made it easier for the affected systems to get the fix that was needed. However, as in the classic Chinese proverb "while the priest climbs a foot, the devil climbs ten", as quickly as the good guys build a wall that is a foot high, the bad guy will try to climb ten feet to get over it. When newer and more sophisticated vulnerability was discovered, more security patches were put out. From the first release of Windows XP in 2001 to the end of support in 2014, low estimate indicates that Microsoft sold more than 500 million copies of the product, making it one of the most successful products in Microsoft history. However, it is also notoriously known as the product with most security patches, though no known figure has been ever released and because many fixes were rolled into a particular update (i.e. fixing multiple problems within a single update), the final figure will never been known. As of today, even though the official technical support for the product has been ceased, but continued use of this version of windows software in many computers (e.g. Automated Teller Machines in banking industry) and platforms and existing versions of other software running in this windows operating system necessitates unofficial support, either by the company or by third party security vendor for a long time to come.

While it is successful as a product, it is a disaster in from cyber security perspective. What went wrong? A number of factors likely attributed to the problems, to name a few:

- Functional design over security design
- Proliferation of Internet
- Being a big target
- Quick to market

10.2.1 Functional Design over Security Design

The number one goal of any product is design is to ensure that the product works as intended. As a graphic user interface and operating system to the ubiquitous personal computer, it has been a fantastic product, making millions and millions of users' lives easier in simple and effective use of their personal computers, providing intuitive learning curve that easily let novice to master the complex operation of the personal computer. As a security tool, depending on from whose point of view, it was disaster for many corporate IT security departments but a great platform to explore, experiment, and crack for the hacker community.

It was inconceivable that with the tools, know-how, security expertise and vast resources of Microsoft that security of Windows XP is an after thought and not an integral part of initial design. However, given the vast majority of the patches are so-called "hot" fixes or patches, it was clear that somewhere during the product development cycle, security has been overlooked, possibly at the design stage, development stage or even at the testing stage. While it may not have been intentional, the effect is all the same. Tasks and processes that could have prevented some of the vulnerabilities and flaws at design stage are establishing a threat model, performing a thorough risk assessment and threat analysis, as well as analysis of capabilities of hackers. During product development stage, adhering to secure coding practices, moving away from obscurity, and adding security controls can make a huge difference. In the testing stage, performing thorough penetration testing, extreme boundary testing, and removing test codes are also good practices to follow.

10.2.2 Proliferation of Internet

The explosion of the Internet age has made information readily available, not just for the good, but also for the evil. While more and more corporate networks are being connected to the Internet, more and more personal computers are also being connected to the Internet, making hacking possible in every corner of the globe for as long as there is access to the Internet. Known vulnerabilities can be spreading almost faster than the speed of light through online hacking communities such as 2600, an underground hacking community that traced its roots to the early 1980's. Internet also made sharing of information easier and faster. A new age of hacking also exploded. Powerful personal computers made it easier to take the software to its limits, the limits that were not easily discovered during extreme boundary testing, and discover new bugs and vulnerability that had not been taken into account.

10.2.3 Being a Big Target

At one point, Microsoft was the biggest private company that rode the growth of personal computers to fame, money and glory. Being big also gets the company on the radars of many hackers and the entire hacking community. Earlier exploitation of the software vulnerability was no doubt motivated by financial gain, for example, hacks that led to using the software without paying for the licensing of the software. Later exploitations were motivated by bigger and more lucrative potential of getting into the corporate networks, stealing corporate secrets and even trading or selling them for illegal financial gains. Unbound by the ethics, these hackers pooled their resources together: reverse engineered, shared hacking tools, distributed computing power, and published their work so that vulnerabilities can be exploited to the fullest. Unfortunately for the company, before the company can react to many of the newly discovered security holes, the attackers and hackers tried their best to take full advantage to their gains. By the time old holes are plugged in, new holes have been discovered. It became a cat-and-mouse game between the giant corporation and the seemingly invisible hacking community that hot fixes and patches have become the norm rather than exception where the bad guy always seemed to be one step ahead. According to the latest data release by CVE Details, a online database that tracks vulnerabilities by vendor, product and by version, the number of vulnerabilities for Windows XP totalled more than 700 from 2000 to 2014 where the data was collected.

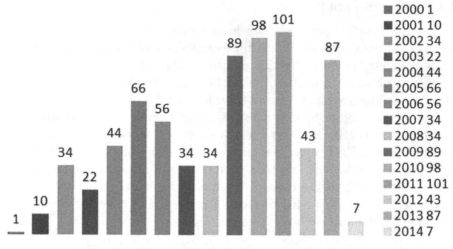

Figure 10.1   Vulnerability by year.

10.2.4 Quick to Market

Even when Microsoft was the big company at the time, it did not stop them from releasing products available to the market as quick as they can as to grow the company market share and make investors happy. While it is difficult to tell whether shortcuts have been taken as the pressure mounted to release the product by a certain time and updates to the release yet another time frame, it is extremely likely that the rushing of product to the market place attributed to being less thorough in other phases of the product development cycle, for example, testing done in haste. This further shows that security cannot be rushed.

By the time Microsoft stopped supporting Windows XP, millions of users have become cyber security victims from using the Windows XP and the numbers of applications supported by the software itself either directly as a result of the vulnerability of the software or indirectly as a result of virus and malware that benefited from the insufficiencies of the security of the software. For better or for worse, this incidentally has spawned an entire industry of security and anti-virus software that proliferated during the period. The losses from productivity, corporate security incidents handling and recovery, development of anti-software, theft of personal information, etc., are far greater and are much more difficult to quantify. This was indeed an expensive lesson learned by everyone.

10.2.5 Design Aspect

While security by design is definitely not something new, it has been tradition-ally a principle and practice followed in the software development community where security is designed from the ground up. Security by design can be many things to many people, but in general, it consists of many good design principles and industry best practices:

- Design from ground up, not as an add-on
- Not relying on secrecy
- Adhering to the principle of least privilege
- Isolation of systems and components
- Defence in depths

These easy to understand principles and guidelines can be generalized to put into practice in all industries, from IT to telecom to the more security-conscientious Internet industries and eliminate the security holes before they are built into the products. If these are principles so simple to understand and follow, why are all products not built this way? This question begs to be answered. As seen previously, some of the reasons for not adhering to and following through with the principles and guidelines of security by design have many factors. Some of the factors that impact the imple-mentation and design decisions to overlook security in favour of strictly adhering to the guiding security principles include faster-to-market pressure of beating competitor with new (sometimes improved) products and services and constantly meeting market demand that trumps a fundamental sound design, lack of understanding of security, and the lack qualified security professionals make security decisions difficult for the design team and for the management team that has the ultimate decision of go-no-go. The old telecom carrier's mentality of "I can live with that" from the financial perspective that essentially says when the potential loss to be incurred or expected due to attacks from cyber security breaches is less the cost than the controls and capabilities needed to combat cyber security, it is simply a part of cost of doing business. Relying on secrecy resulted in many unpleasant incidences that made news stories both entertaining and scary, especially after what the former CIA contractor Eric Snowden revealed and what a certain Internet router vendor revealed. Those revelations certainly stroke quite a few nerves.

When the users of the product and services are impacted, their privacy is on the line and therefore privacy concerns are also increasing impacting the ground-up approach to security by design. Together with security by design, privacy by design is yet another powerful approach to enhance system security by taking privacy into account in the early stage as well as through the entire design and engineering process. This concept originated by a joint team of Information and Privacy Commissioner of Ontario in Canada, the Dutch Data Protection Authority, and the Netherlands Organization of Applied Science in 1995 offered some seven foundation principles:

1. Proactive not reactive; Preventative not remedial
2. Privacy as the default setting
3. Privacy embedded into design
4. Full functionality – positive-sum, not zero-sum
5. End-to-end security – full lifecycle protection
6. Visibility and transparency – keep it open
7. Respect for user privacy – keep it user-centric

However, while the goals of secure design are common, privacy by design is mainly driven by the various regulations that are sometimes difficult to understand and contradicting. Privacy regulations and to some extent are tightly coupled with data protection regulations are highly regionally in nature and differ greatly from region to region and from country to country. Keeping track of the numerous regulations and coming up with a one-size-fits-all design to satisfy all of the regulations can be an enormous undertaking for anyone, even for large corporations that are dictated by the regulations in regions and states where products and services are marketed. Throwing additional complexity to the equation is the fact that many counties' civil and penal codes have not been updated in years to keep up with the growth of Internet, communication, and privacy protection for individuals making penalty for not complying (intentionally or unintentionally) a mere slap on the wrist.

10.3 Cyber Security Paradigm Shift

While more focus and efforts have been put into applying the principle of security by design to practice, the fact that the concept has been around for a long time and yet the number of cyber security incidents is still on the rising means that the entire industry is still trying to win the battle against the hacking community. What if a vendor can have a guarantee that the product is secure? What if the vendor can provide assurance and proof that the product is secure against all threats and vulnerabilities at the time that the product is being released? That concept already exists but applies only to a number of limited systems that require high security (e.g. military-grade systems and networks). But if properly and universally applied to all system and network design, engineering, and building, it can improve security dramatically. The process to make this happen is called "security assurance" and is being taken by storm in the telecommunication industry as a paradigm shift. It is a combination of good security by design, security by proof, and security by assurance that makes this approach effective.

10.3.1 Security Assurance

The market needs and lacking of a security assurance (provable security) for telecommunication sector coupled with the explosion of mobile broadband growth have made the condition ripe to start putting the focus on an industry-wide security assurance process for telecommunication products. Attacks have become more and more sophisticated; attackers have become more and more intelligent; and the attack tools have become more and more advanced. At the same time, the wireless networks have become more and more open in terms of architecture and visibility. The combination of these events has called upon the entire telecommunication community to re-evaluate the approach and the process to ensure that the products are more and more secure. This leads to a need for developing an international security assurance standards for the wireless telecom product so that one process can be applied to every market, every product and for every stakeholder. As a result, expert members in the Third Generation Partnership Program (3GPP) security group partnering with Group Speciale Mobile Association (GSMA) have committed themselves to create such a security assurance process.

10.3.2 Security Assurance Challenges

Riding the wave of 3G, 4G and the forthcoming 5G, operators have introduced a plethora of new services that not only rely on new products and features to be developed quicker than ever before but also that these products and features can be rolled out to the market place at a record pace. With a great awareness of the factors that attributed to the problems and pitfalls of earlier cyber security incidents as the network security landscape and threat models are constantly evolving, the users and the operators demand and expect the greater security and more values they have become accustomed to and offered by the vendors. The challenges are real. The interests are high for customers, governments, and vendors alike to ensure that the telecommunication products and the networks are more secure than ever.

Today's telecommunication networks have become more open, and at the same time more sophisticated and more intelligent. We are relying on the communication networks and connectivity more than ever. Information, tools, and knowledge about networks and network security are readily available to anyone who has the desire and determination to learn about anything or gain a great deal from it (i.e. proliferation of the Internet), including those who attempt to seek financial gains (e.g. not paying for the product) or those who seek to create damage and disruption. When the information and knowledge are in the wrong hands, along with the more powerful machines and tools of today (e.g. PCs, tablets, smart phones, etc.), it is becoming increasingly evident that malicious misuse of the learned or gained knowledge can lead to serious disruption of communication networks and network services. Being able to communicate anytime and anywhere also means that attackers are able to launch attacks anytime and anywhere. The potential losses can be great in terms of productivity losses, financial losses and information losses. It has become vitally important to keep the communication systems and network more secure than ever. Ensuring the security of systems and networks has become one of the toughest challenges for the entire telecom ecosystem in the foreseeable future as well as the whole ICT industry. Some of these challenges include:

- Identity management
- Virus, malware, worms, and botnets
- Internet-based attacks
- Industrial espionage and sabotage
- Privacy, data, and cyber security laws, directives, and regulations
- Awareness, education, and training

These challenges also bring along an assortment of many potential threats and risks to the products, networks, and services:

- Physical tampering
- Denial of service and attacks on the networks
- Compromise of authentication credentials
- Man-in-the-middle attacks on the networks
- Intercepting and modification of user's data
- Rogue network equipment
- Mis-configuration
- Radio resource tampering
- Hackers

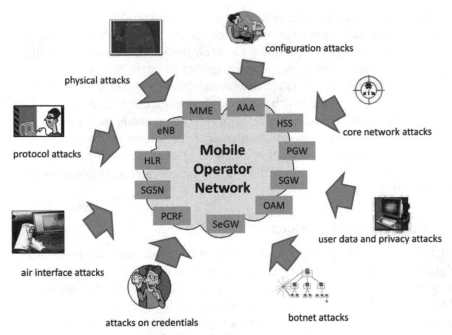

Figure 10.2 Threat and attack model.

10.3.3 Market Place Challenges

In addition to building secure products, meeting these challenges also requires that a global security assurance process to be developed so that the security of the products can be demonstrated in a systematic approach. The needs for such security assurance are also driven by the market for such a transparent process, such as in the Indian market and other markets around many parts of the world. Making claims that a particular product is either secure or insecure without substantiating the claims is simply irresponsible. From the operators' perspective, they demand that the products they place in their networks are secure and are developed with the highest integrity as they not only have their reputation to protect, the user's security and privacy to protect, but also the laws and regulations to comply with as there are numerous such regional and national laws and regulations regarding the protection of user data and user privacy. From the vendor's perspective, they want to ensure that not only the operator's requirements are met and that their products are secure, but also that they can keep up with the demand for faster feature and product development without compromise in security as they too, have a reputation to protect. At the same time, the vendors also want an environment to ensure that their products can be used in all markets of the globe without the need to customize and certify the products for each market. To reach this goal, the operators and the vendors have come together in the telecommunication industry create such a security assurance standards to be applied, recognized, and accepted in all areas for which network products are sold and marketed. The bottom line is that the telecommunication industry needs such a security assurance process so that every stakeholder can benefit from it. Products built to the security assurance specification and having gone through the security assurance process will be able to withstand any unsubstantiated claims about the security of the products.

10.3.4 Regulatory Challenges

Regulators have spoken loudly and clearly with cyber security laws and regulations put in place in various regions around the world covering data security, user privacy, and with even stricter requirements for telecommunication products when the systems and networks provide vital services to serve the government and regulator communities in addition to serve the general public. Some examples of these regulations and directives include EU's Data Protection Directive, UK's Data Protection Act, Canada's Privacy Act, Japan's Personal Information Privacy, China's Provisions for Administration of Information Service, China's Cyber Security Laws, so on and so forth. The list can get quite lengthy. It should be noted that the various regulators should not overly burden the industry with unnecessary and impractical rules and regulations that are difficult to harmonize and thereby creating a barrier to inhibit innovation. Nevertheless, laws and regulations are in place for good reasons and offer reasonable amount of deterrence.

10.3.5 Requirements of Security Assurance

Meeting the challenges head-on requires that the security assurance process must be transparent, collaborative, global, standards-based, and practical in order to be effective. Requirements from operators, vendors, and regulators need to be taken into account fully to build up the security assurance process. Market needs drive products and services, which in turn drive requirements on the vendors, operators and regulators. The market place is unyieldingly unwilling to compromise even in the face of increased threats and risks presented. Regulatory requirements are also an important and necessary component of the security assurance process. All and all, it requires the best of vendors, operators, and regulators to come together to define ways of ensuring the security of products, systems, networks, and the users. The network operators also have requirements and demand the best of their vendors and suppliers not only to build the best and most secure products, but also to provide indisputable evidence or secure proof, in the form of assurance and certification. The operators have moral, legal, and financial requirements and obligations to ensure the security and privacy of its customers – the very users who contribute to the growth and success of the operators. Besides building the most secure product, vendors and suppliers are also required to ensure that not only they follow the most strict industry security best practices, but also receive the necessary accreditation to ensure that they are held to the highest standards.

10.4 Security Assurance Process

Strictly speaking, the security assurance process is certainly not something new. The process already exists for many reasons for IT products. The process can be used for networking products to a certain extent but with some degree of limitations. The most notable of these developments is the Common Criteria (CC) and the Common Criteria Recognition Arrangement (CCRA) that were based on combination of three standards: Information Technology Security Evaluation Criteria (ITSEC), Canadian Trusted Computer Product Evaluation Criteria (CTCPEC), and Trusted Computer Security Evaluation Criteria (TCSEC) originally developed by the governments of European countries, Canada, and US respectively. Because of the cost of certifying a particular product or a system, that process has been used highly secure systems and networks such governmental networks and military systems.

10.4.1 Goals of Security Assurance

In recent years, some of the legal requirements have put the vendors and operators in a dilemma as they risk the possibility of financial losses in terms of fines originating from government enforced penalties and lost business opportunities as result of negative publicity when these requirements are not met. Vendors have to show not only that they have followed the strict operator requirements and legal requirements, but also industry best practices that they have built the products to the highest degree of standards, but also with highest degree of security and integrity. Furthermore, vendors may have to repeat the process in every market and region where their products are deployed.

There should be a globally standardized process in demonstrating the security of products, systems, and networks. This process has to be systematic that every stakeholder can work with and rely upon. The stakeholders include vendors, suppliers, network operators, regulators, government agencies, etc. With so much at stake, it is easy to understand the goal of this approach – to specify the network product security assurance requirements that are necessary to protect against unwanted access to the product, its operating systems, and running applications. The security assurance requirements to be developed and specified should be based on threat and attacker models that are applicable to the functions the products are designed to perform, including generic IT and communication functions. The security assurance requirements are of course in addition to any basic functional requirements and feature requirements of the products to be developed. For instance, a base station will be developed with set of required core functions (e.g. RF, communication, etc.) with the security assurance requirements in mind as these security assurance requirements are taken as baseline for building products that are not only functional, but also demonstrably secure. This systemic approach becomes the "Security Assurance Process".

10.4.2 Challenges of Security Assurance

Although CC and CCRA have existed for many years and have gained international acceptance with more than twenty member countries around the world, but the framework and infrastructure were developed mainly to focus on IT products as well as computer products, and were originally developed to serve the government and intelligence markets.

Many attempts to apply the CC and the CCRA framework for certain telecommunication products have shown the process to be both intensive and time consuming and may not meet the need for products to reach market timely with many of the features and products that are required to offer value added services to the users. Obtaining CC certification for products, even at particular levels acceptable to private communication networks (e.g. EAL 3) would mean thousands of dollars and many more man-hours spent with one of the CC-accredited laboratories around the world before certification can be obtained. Obtaining certification for higher level of assurance requires even greater efforts, more time and more money. To that effect, CC has been around many years and has served a good purpose. Many members within the telecommunication community felt that it does not address the constant changing needs and requirements of the telecommunication industry since it was designed for a different class of products and it may prove difficult to adopt it to accommodate the security assurance requirements of telecommunication products without substantial modifications. Trying to fit the networking products into the CC framework for the purpose of security assurance and accreditation has proven to be awkward, time-consuming, and expensive even though CC is not without its merits. The lessons learned and experience gained through CC and the CCRA framework will serve as a solid basis for developing security assurance process in the telecommunication environment.

Another challenge is having the stakeholders to endorse the process once it is completed. For CC and CCRA, it took quite some time for it to be recognized in twenty-plus countries, mostly through governmental efforts. Without these efforts and relying on the industry leverage alone may prove difficult even though 3GPP is an internationally recognized standards body which produces de facto specification for wireless systems around the world. Recognition of 3GPP standards is often quite different than recognition of security assurance standards such as interoperability aspects.

Yet another challenge is choosing the right threat model and security framework in the process. There are various threat models and security framework created for different purposes, such as STIX, short for Structured Threat Information Expression, STRIDE, short for Spoofing, Tampering, Reputation, Information Disclosure, Denial of Service, Elevation of Privilege, or ITU's X.805 Security Architecture for systems providing end-to-end communications. Though there are similarities among them, but like CC, they were developed for specific cases making adaptation in telecommunication difficult. There is no right or wrong threat model and everyone has its own merits.

10.4.3 3GPP Security Assurance

A good example of putting the lessons learned through CC and the CCRA framework to use is shown by 3GPP in its Security Assurance Methodology (SECAM) and Security Assurance Specifications (SCAS) activities. This work is also done in conjunction with GSMA's Network Equipment Security Assurance Group (NESAG) (In 2015, NESAG has been subsequently renamed Security Assurance Group or SECAG for short) where as 3GPP defines the security baseline specifications including test cases for evaluating the results while SECAG defines the framework for accrediting evaluation laboratories (including vendor evaluation laboratories and third-party evaluation laboratories) and resolution in case of disputes between the vendors and operators.

SECAM and SCAS are seen as a positive development of such a security assurance methodology specifically for the 3GPP products, as the first attempt to evolve into an international standards purely from the industry perspective. SECAM and SCAS are intended to be a comprehensive process for which all network product and network product when the process is fully implemented. It starts with identification of the threats and risks associated with each product. Although there may be many functions within each product (e.g. encryption, authentication, etc.), the focus is to perform the threat and risk analysis on the entire product as a whole so that the security requirements along with security assurance specification can be developed. Next, the security requirements are developed for that product, which may be done in modular fashion, for example based on functional components, to afford the flexibility of applying these modules of requirements to different products with same or similar functions without duplicating efforts needed to develop security assurance requirements for the same function in another product.

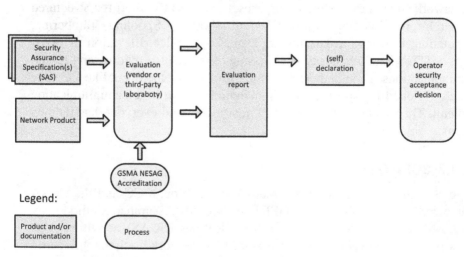

Figure 10.3 3GPP security assurance process.

The goal of the security requirements are to take into account threat model and risk analysis, attacker potentials and capabilities, environmental variations, etc. to resist all known attacks (both current and anticipated). A testing process (e.g. evaluation by an accredited laboratory that can be either a vendor or a third-party independent laboratory) then follows with test cases and testing methodology that will be able to produce verifiable and repeatable results from the security testing of the product. All products are to be tested vigorously, thoroughly, and comprehensively to ensure that that the product being built and tested will conform to the security requirements. This is the same philosophy of assuming nothing, believing in no one, but checking everything with a multi-layered "many hands" and "many eyes" approach to independent verification in order to reduce the risk of insecure products being developed, produced, and eventually deployed.

The final step in the security assurance process is for the operator to gain confidence after the product has clearly demonstrated meeting all of the defined security requirements and passing all of the verifications. Acceptance is also backed by rigorous and yet robust audit mechanism where verifiability, traceability, and disputes can be resolved.

More than ever, the operators demand greater uniformity in terms of requiring the same security assurance from all vendors, especially in a multi-vendor deployment environment. The users demand unequivocal guarantee in terms of security and privacy. And finally the governments have placed stringent requirements on the service providers to deploy secure network equipments and networks in the name of national security as more and more governments are relying less on dedicated private networks and more on public networks to carry their traffic. Greater emphases are placed on the security assurance of telecom products, as the entire industry has experienced an evolution of migrating to open architecture and open platform from a traditionally considered closed environment. This, in turn will drive the telecommunication community to create the standards necessary to achieve the desired security assurance for the products developed to comply with specifications.

10.4.4 3GPP Security Assurance Approach

In order for the security assurance process to work, a multi-layered approach needs to be taken where the security assurance process is to be developed as an open and transparent process where all stakeholders with their vast expertise and experience from aspects of telecommunication, information technology, networking, and security contribute toward a common goal. The ultimate goal is not only to give the operators assurance that the products built are secure, but also to give assurance to all stakeholders that the products built are secure against the known attacks at the time of deployment.

Every stakeholder across all regions also needs to sign on to and agree to the process. Working with other standardization bodies, such as ETSI, IETF, ITU, 3GPP, etc and with regulators will ensure mutual acceptance once a particular product has successfully gone through the security assurance process is not alone in perfecting the security assurance process. Other organizations' expertise is also very helpful as those have gone through similar efforts such as National Institute of Standards and Technology (NIST) Security Content Automation Program (SCAP), the Security Automation and Continuous Monitoring (SACM) in IETF, and SECAM and SCAS in 3GPP even though sometimes the efforts by disjoint organizations would appear in random. Rest assured the seemingly disjoint efforts all contribute to the same goals in their own ways. Mutual recognition and mutual acceptance equate to a single and fair process for all and goes a long way to ensure the success of the entire process.

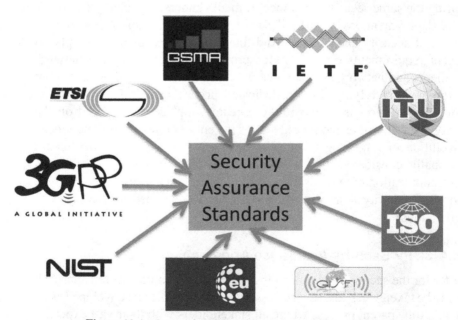

Figure 10.4 International security assurance collaboration.

Though the aim is for the operators to gain confidence about the security of the products, but since the process is open and every step within the process can be documented and substantiated, it should make it very easy for all stakeholders, whether they are vendors, system operators, regulators or the like to realize the transparency within the process and to accept the products with a great deal of confidence that the products are secure, once the rigorous security assurance process has been followed. It is important to emphasize once again that such a security assurance must be transparent, collaborative, global, standards-based and practical to be effective. It is noted that as a standards-developing organization (SDO), 3GPP has made great strides to achieving the goal of security assurance through the study items and work items that have produced various technical reports such as 3GPP Study on Security Assurance Methodology for 3GPP Network Products Technical Report, 3GPP Pilot development of Security Assurance Specification (SCAS) for MME network product class Technical Report, and 3GPP Security Assurance Methodology for 3GPP network products Technical Report. Once the work is done, 3GPP will have produced security assurance specification for all 3GPP products, starting with the Mobility Management Entity and extending to other network product class such as evolved NodeB (eNB), Packet Data Gateways (P-GW), etc.

In summary, here is the approach and process taken in 3GPP to develop a security assurance specification:

1. Establish and methodology for 3GPP security assurance
2. Create testable requirements and test cases
3. Develop specification on security assurance for a particular 3GPP product (e.g. MME, P-GW, eNB, etc.)
4. Test the process
5. Extend the process for all 3GPP products.

10.4.5 Security Assurance around the Globe

Recognizing the importance of security assurance in ICT, many other SDOs have accelerated their pace in bringing more awareness through the development and work on security assurances. Work in CC is continuing in terms of making it more "user-friendly" by considering the different levels of assurance. IETF has taken on the work of creating use cases for Endpoint Security Posture Assessment and has gone through several iterations of internet drafts. GISFI is working closely with 3GPP in addressing the security assurance requirements originating from the Indian telecom market. CESG, though not a SDO, has also development a commercial product scheme or CPA for short for the UK market to provide assured commercial security products for users who have a need for information assurance. With network virtualization and SDN on everyone's mind, the European Union Agency for Network and Information Security has also developed a set of assurance criteria to assess the risks of adopting cloud services, obtain assurance from the cloud providers, and reduce the assurance burden on the cloud providers. Similar activities in ITU produced Entity Authentication Assurance Framework in ITU-T's X.1254. The list goes on and on, but none bigger than the challenges taken up in 3GPP community as seen in the initiative, commitment, and paradigm shift to the security assurance process.

10.5 Conclusion

Now is the time to step up and address future global cyber security challenges today. The team of security experts representing the operators, vendors, and regulators have come together in telecommunication community (e.g. 3GPP) in unison to focus on the security assurance process by leveraging the expertise and knowledge gained from years of creating standards. Making claims one way or another about the security of the products, systems, or networks based on misconstrued or misinformed belief without any concrete evidence is not only irresponsible but they are also not valid reasons to accept or reject a particular product. The systems and networks of tomorrow start today. Getting away from the cycles of break-and-fix model to drive the security assurance message home will ensure the success of verified and certifiable security claims as well as provably secure system and network design process. Security begins with a commitment coupled with a solid foundation of understanding the threats, defining a security assurance process, and going through vigorous testing leading to verification. It is also important for the market that a successful verification as a result of the security assurance process should be internationally recognized and accepted as the process itself is an open process. Following the security assurance process will not only be beneficial for the telecommunication industry, but also a win-win proposition for the entire ecosystem. With the paradigm shift in combating cyber security battles as the entire ecosystem eagerly anticipating the turning of the tide against the hacking community, it is time to start winning.

References

[1] Privacy by Design The 7 Foundational Principles, https://www.ipc.on. ca/wp-content/uploads/Resources/7foundationalprinciples.pdf

[2] Canada's Privacy Act, http://www.priv.gc.ca/leg_c/leg_c_a_e.asp

[3] Cyber Security Laws, Cyberspace Administration of China, http://www. npc.gov.cn/npc/xinwen/2016-11/07/content_2001605.htm

[4] EU Directive 95/46/EC, The Data Protection Directive.

[5] The CC and CEM documents: http://www.commoncriteriaportal.org/cc/

[6] The CCRA introduction: http://www.commoncriteriaportal.org/ccra

[7] CCRA Licensed Laboratories: http://www.commoncriteriaportal.org/ labs/

[8] Common Criteria for Information Technology Security Evaluation, Version 3.1 Release 4, September 2012.

[9] Cloud Computing Information Assurance Framework: http://www.enisa. europa.eu

[10] CESG Commercial Product Assurance (CPA) Scheme: http://www.cesg. gov.uk/servicecatalogue/Product-Assurance/CPA/Pages/CPA.aspx

[11] IETF Internet Draft: "Endpoint Security Posture Assessment – Enterprise Use Cases".

[12] ITU-T X.1254: Entity Authentication Assurance Framework.

[13] 3GPP TR 33.805, Study on Security Assurance Methodology for 3GPP Network Products.

[14] 3GPP TR 33.806, Pilot development of Security Assurance Specification (SCAS) for MME network product class.

[15] 3GPP TR 33.916, Security Assurance Methodology for 3GPP network products.

[16] 3GPP TR 33.926, Security Assurance Specification (SCAS) threats and critical assets in 3GPP network product classes.

[17] 3GPP TS 33.116, Security Assurance Specification (SCAS) for the MME network products class.

[18] 3GPP TS 33.117, Catalogue of general security assurance requirements.

Index

About the Editors

Samant Khajuria is associate professor at the Center for Communication, Media and Information Technologies (CMI) Copenhagen at Aalborg University (Denmark). He earned his Bachelor degree in 2006 in electronics and communication from PES Institute of Technology – Bangalore (India); and Master's Degree in Communication networks (specializing in security) from Aalborg University in 2008. He was enrolled at Electronics Department, Aalborg University in 2009 and graduated in 2012 with a Ph.D. degree. His research Interests are in the areas of Trust and Privacy, Cryptography, Computer and Network Security. Some specific research topics he has addressed in the past include design and cryptanalysis of Single-pass Authenticated Encryption Stream ciphers, crypto solutions for reconfigurable devices, securing Software Defined Radios.

Lene Tolstrup Sørensen is associate professor at CMI (Center for Communication, Media and Information Technologies), Electronic Systems, at Aalborg University Copenhagen. She holds a M.Sc.E.E. from DTU (Technical University of Denmark) and has specialized in Interaction Design hereunder usable privacy and user requirements elicitation for new IT developments. Her main interests are on: Interaction and participative design processes, usable privacy and security, and privacy tools. She has published more than 100 scientific papers and reports.

Knud Erik Skouby is professor and founding director of center for Communication, Media and Information technologies, Aalborg University-Copenhagen. Has a career as a university teacher and within consultancy since 1972. Working areas: *Techno-economic Analyses; Development of mobile/wireless applications and services: Regulation of telecommunications.* Project manager and partner in a number of international, European and Danish research projects. Served on a number of public committees within telecom, IT and broadcasting; as a member of boards of professional societies;

as a member of organizing boards, evaluation committees and as invited speaker on international conferences; published more than 300 scientific papers, books and reports. Editor in chief of Nordic and Baltic Journal of Information and Communication Technologies (NBICT); Board member of the Danish Media Committee. Chair of WGA in Wireless World Research Forum; dep chair IEEE Denmark.

CPSIA information can be obtained
at www.ICGtesting.com
Printed in the USA
LVHW081800160323
741793LV00002B/7

9 788793 519664